T0292276

New Frontiers in the Study of Social Phenomena

Federico Cecconi
Editor

New Frontiers in the Study of Social Phenomena

Cognition, Complexity, Adaptation

Editor
Federico Cecconi
Institute of Cognitive Science and
Technology – CNR
Rome
Italy

ISBN 978-3-319-23936-1 ISBN 978-3-319-23938-5 (eBook)
DOI 10.1007/978-3-319-23938-5

Library of Congress Control Number: 2015959568

Springer Cham Heidelberg New York Dordrecht London

Printed on acid-free paper

Springer International Publishing AG Switzerland is part of Springer Science+Business Media
(www.springer.com)

Contents

Contributors

Giulia Andrighetto LABSS. Laboratory of Agent-Based Social Simulation, Institute of Cognitive Sciences and Technologies – CNR, Rome, Italy

Alessandro Barazzetti QBT, Sagl, Chiasso, Switzerland

Chiara Bassetti Laboratory for Applied Ontology, Institute of Cognitive Sciences and Technologies – CNR, Trento, Italy

Juliana Bernhofer Department of Economics, Ca' Foscari University of Venice, Venice, Italy

Marco Campennì Arizona State University, Tempe, AZ, USA

Federico Cecconi Institute of Cognitive Sciences and Technology – CNR LABSS. Laboratory of Agent-Based Social Simulation, Rome, Italy

Rosaria Conte LABSS. Laboratory of Agent-Based Social Simulation, Institute of Cognitive Sciences and Technologies – CNR, Rome, Italy; Institute of Cognitive Sciences and Technologies, CNR LABSS. Laboratory of Agent-Based Social Simulation, Rome, Italy

Giuliana Gerace Department of Philosophy, University of Pavia, Pavia, Italy

Francesca Giardini Laboratory of Agent-Based Social Simulation, Institute of Cognitive Sciences and Technologies – CNR, Rome, Italy

Rosangela Mastronardi QBT, Sagl, Chiasso, Switzerland

Luis G. Nardin LABSS. Laboratory of Agent-Based Social Simulation, Institute of Cognitive Sciences and Technologies – CNR, Rome, Italy

Mario Paolucci Laboratory of Agent-Based Social Simulation, Institute of Cognitive Sciences and Technologies – CNR, Rome, Italy; Institute of Cognitive Sciences and Technologies, CNR LABSS. Laboratory of Agent-Based Social Simulation, Rome, Italy

Daniele Porello Laboratory for Applied Ontology, Institute of Cognitive Sciences and Technologies (ISTC-CNR), Trento, Italy

Valentina Punzo Department of Law, Society and Sport, University of Palermo, Palermo, Italy

Aron Székely LABSS. Laboratory of Agent-Based Social Simulation, Institute of Cognitive Sciences and Technologies – CNR, Rome, Italy

Daniele Vilone Laboratory of Agent-Based Social Simulation, Institute of Cognitive Sciences and Technologies – CNR, Rome, Italy

Chapter 1
Computational Social and Behavioral Science

Rosaria Conte

Computational Social Science (CSS) is a novel research field that exists at the intersection of social science, complexity and Information and communications technology (ICT) science. Born as a new scientific instrument for modelling social phenomena, CSS dates back to the seventies. Its appearance coincided with the first appearance of World Models as a result of applications of System Dynamics to the study of demographic development and planet anthropization. In 1976, World Models received the attention of politicians and statesmen in the Club of Rome, which paved the way to European unification a long time before the realization of the European Union. Though wrong, the World Models theory significantly contributed to scientific innovation in major fields among various disciplinary areas.

Recently, the advent of Data Science, especially Big Data, gave rise to CSS. Big Data refers to the vast heterogeneous datasets compiled by users of ICT technologies. The wealth of new data demanded a quantitative variant of CSS. Unlike CSS modeling, quantitative CSS benefits from computers' ancillary role (cf. Conte and Giardini 2014). Computers are increasingly being used as vectors and repositories of empirical data, and as instruments of social data mining (or simply social mining). As we shall see, the quantitative variant was launched at the expense of the modeling variant of CSS, which not only produces tools but also and a new way of modeling.

R. Conte (✉)
LABSS. Laboratory of Agent-Based Social Simulation, Institute of Cognitive Sciences and Technologies – CNR, Via Palestro 32, 00185 Rome, Italy
e-mail: rosaria.conte@istc.cnr.it

F. Cecconi (ed.), *New Frontiers in the Study of Social Phenomena*,
DOI 10.1007/978-3-319-23938-5_1

1.1 Social and Natural Sciences: An Uncertain Alliance

There is no doubt that CSS represents a great opportunity for the renaissance of the social sciences, a discipline that started to decline after the great successes of big thinkers like Durkheim, Comte, Pareto, Freud, Piaget, Keynes, Vygotsky, and others between the nineteenth and the twentieth centuries.

Part of the reason for the decline of the social sciences was the result of the romantic reaction against enlightenment, which caused a profound chasm between the so-called sciences of the spirit and the sciences of nature. Many representatives of the storicistic movement (see, for example, the Italian philosopher, Benedetto Croce 1917) and of the hermeneutic philosophy (from Friederich Schleiermacher (1800) to Dilthey (1924)) insisted on the impossibility of subjecting historical and philosophical understanding to the universal laws of physics. To exalt their humanistic aspirations, romanticists and their followers erected barriers around themselves, dispensing at once with physical laws, mathematical formalisms, and experimental methods.

The positivist rescue was not awaited long in coming. Sociologists and economists attained success at the end of the nineteenth century (e.g., Auguste Comte and Vilfredo Pareto) thanks to the import of mathematical models. Though viable instruments, mathematical models did not always bear positive consequences for the development of the social sciences (see Conte et al. 2015). What is worse, they often contributed to creating gaps and barriers between humanists and social scientists. Rather than creating appropriate instruments for investigating social phenomena in a rigorous and controllable way, social positivists imported existing ones that were not always fit for such purposes. On the other hand, humanists did, but often without submitting their research results to intersubjective control.

Social scientists often align themselves at either extreme: isolation or cultural subordination to physical sciences. The limitations resulting from isolation are evident: good science always needs confrontation. Those resulting from cultural subordination, however, require clarification. In particular, why not simply accept the hegemonial role of the hard sciences, and take advantage of their successful results? At first glance, this appears to be a reasonable program. Except that in the end it leads to dispensing with one of the major tasks of science, i.e., understanding the processes generating the phenomena to be explained. Mathematical models applied to social theories do not, and cannot, express the behavioral mechanisms by which social phenomena are generated. In particular, mathematical formalisms cannot express behavioral mechanisms as they are hypothesized to be represented within the behaving systems. A social action ought to be generated by modeling the internal—namely, the mental—mechanisms in terms of mental states and the operations accomplished on them, such as social emotions and reasoning, social learning, and social or collective decision-making, etc. Thanks to the gap between humanists and social scientists, the behavioral mechanisms of social phenomena were overlooked in the positivist study of society and economy, at least until the advent of the modeling variant of CSS. Agent-based modeling and simulation allowed researchers to generate macro-social effects from interaction at the micro-social level. One of the lessons from the humanists was then vindicated, as the algorithms forming part of the computational agent models enable the programmer to express and operate on some proxies of mental representations and operations.

Chances are, however, that the opportunity provided by the modeling variant of CSS will be missed if the whole point of the quantitative variant is to work out new statistical techniques, or even worse, if the quantitative variant of CSS were to definitely prevail over the modeling variant. It is by no means clear that the emergent product of CSS applications is crowd-thinking, or some sort of collective intelligence. A bit of recent history and some current data seem to point us to a potential abuse of social mining techniques, i.e., the practice of social mining (a) in a fragmentary, rather than integrated, way; (b) in a commercial/speculative-oriented rather than governance-oriented context; and (c) with a focus on short-term forecasting rather than on policy modeling. Social mining abuse, rather than promoting crowd-thinking, could favor instead crowd-faking, or crowd-pollution. Even when based on truthful information, forecasting can have toxic effects if no collective control on the use of acquired knowledge is applied. Toxic effects can make the social environment more confused, atomized, and competitive, because they reduce predictability, resilience, and cooperation. Let us see why.

1.2 A Bit of History

In 2001 the National Academy of Science organized the Sackler Colloquium on the future of the social sciences at Irvine, California, gathering scholars from all over the world and then publishing the proceedings on Proceedings of the National Academy of Sciences of the United States of America (PNAS). Agent-Based Modeling (ABM) was anointed, at what can legitimately be considered one of the most prestigious international scientific institutions, as the leading field in the social sciences for the successive five years at least. This consecration of ABM was due to dissatisfaction with the assumptions of rational choice theory and the consequent necessity for grounding social theories on new interdisciplinary, operational, and falsifiable foundations.

Unfortunately, ABM did not keep some of its promises (Conte e Paolucci 2014). For one example, think of Gummerman and colleagues' project on the extinction of Anasazi, who inhabited the Long Valley until the twelfth century and suddenly disappeared for no apparent reason. Despite generous funding by the Brooking Institute, the project did not yield novel results.

Furthermore, a many of the models worked out through ABM are based on assumptions similar to the implausible postulates of classical game theory (like that of complete and transitive preferences). When not based on rationality assumptions, agent-based models are either too simple to be applied to interesting phenomena or too complex to deliver robust and reliable results. Usually, ABM works better for constructing abstract theories than for modeling real-world phenomena such as the Anasazi's extinction. Nor does ABM work any better at predicting real-world events.

The economist Brian Arthur, in a famous Santa Fe Working Paper with the promising title "Out-of-Equilibrium Economics and Agent-Based Modeling," gave a brilliant explanation of the application problems of ABM. He found one characteristic of complex agents like humans—i.e., the propensity to act based on expectations of future events—to make useful predictions difficult. Arthur defined the inhibitory effect of anticipation as paradoxical, and called it the paradox of anticipation. As an example, it is difficult to enjoy the best pub in the neighborhood in Santa Fe (the El Farol Bar) because everybody will make a guess about which the week night the pub is less crowded. But since all will doing the same thing, the more widespread an accurate prediction is, the less likely it is to succeed: all customers will find the pub as crowded as usual, but on a different night of the week. A similar example is smart holidays. To find the right departure time for the summer holidays is no easy job. Is it better to leave on, say, August 1st or July the 31st? In the early morning or during the night? The same question arises in economic domains, such as the real estate market or the stock market.

In the aforementioned paper, Brian Arthur showed that the more frequent and faster the learning algorithm—i.e., the capacity to learn from ones own and others' observed experience— was implemented, the less successful the simulation and the more chaotic its results. Equilibrium needs time, a luxury we cannot always afford. In the absence of time and coordination, anticipation can be counterproductive—a lesson not so different from Keynes's recommendations in the *General Treatise* (1936), when he warned that coordination should be realized by central institutions. Local, uncoordinated anticipations cause collective emergent effects to worsen. In 2005, Brian Arthur's model did not raise a general alarm. The Western economy had not yet gone through a major economic crash.

1.3 The Present

A few years after the Sackler Colloquium, people started to panic and a public outcry arose. Some blamed the uselessness of econometric models.

It is perhaps of some interest to note what, in the aftermath of the economic crash, a figure of international prestige like Jean-Claude Trichet declared in his opening address when assuming at the Presidency of the Central European Bank, on November 18, 2010:

> The key lesson I would draw from our experience is the danger of relying on a single tool, methodology or paradigm. Policy-makers need to have input from various theoretical perspectives and from a range of empirical approaches (…) we need to develop complementary tools to improve the robustness of our overall framework. Which lines of extension are most promising? Let me mention … avenues that I think may have been neglected by the existing literature (…) we have to think about how to characterise the homo economicus at the heart of any model. The atomistic, optimising agents underlying existing models do not capture behaviour during a crisis period. We need to deal better with heterogeneity across agents and the interaction among those heterogeneous agents. We need to entertain alternative motivations for economic choices. Behavioural economics draws on psychology to

explain decisions made in crisis circumstances. Agent-based modelling dispenses with the optimisation assumption and allows for more complex interactions between agents. Such approaches are worthy of our attention.

Trichet identified ABM as the scientific key to dealing with the crash under the direction of central institutions. His recommendation appeared to be a major advance on the classic view of *homo economicus*, opening up the concept to other disciplines, especially psychology, and to more plausible and useful assumptions concerning decision mechanisms.

However, another part of the scientific community followed a different route. In 2009, a position paper, entitled "Computational Social Science" appeared in *Science*, co-authored by a number of "big thinkers" (David Lazer, Alex Pentland, Lada Adamic, Sinan Aral, Albert-László Barabási, Devon Brewer, Nicholas Christakis, Noshir Contractor, James Fowler, Myron Gutmann, Tony Jebara, Gary King, Michael Macy, Deb Roy, and Marshall Van Alstyne), mostly physicists or ICT scientists. In the paper, presented as a sort of manifesto, a new quantitative, computer-aided science of society was put forward. This publication is generally regarded as the birth of quantitative CSS, which aimed to apply physical-statistical models to the analysis of vast datasets in order to extract significant correlations and favour by this means the anticipation of criticalities.

In 2011, the Proceedings of a Symposium held at Harvard University on the "Hard Problems of Social Science" appeared in *Nature*. Twelve prestigious social scientists— including sociologists, psychologists, political scientists, and economists—converged on a top-ten list of the hardest problems in the humanities and the social sciences. Strangely enough, among the concerns of the Harvard scientists, the prediction of an economic crash did not appear, or it was perceived as less important than the efficacy of social influence oriented to public-utility objectives. Incidentally, Christakis and Fowler—who, together with Gary King and Michael Macy, represented the social sciences in the *Science* paper—also participated also in the Harvard symposium. But this time they expressed quite another opinion. In a social-scientific context, they preferred to emphasize the Trichet necessity for new and efficacious policy instruments in order to understand how to make people quit bad habits—a problem that economists like Pigou unsuccessfully attempted to solve by means of economic incentives (pigouvian tax). This inconsistency by Christakis and Fowler should not come as a surprise: the cultural subordination of some social scientists to the physical sciences is not new. But in this case, with all due respect to Jean-Claude Trichet and Brian Arthur, it the result is to convert CSS in social mining.

In the meantime, in Europe, the FuturICT project participated in the race to secure FET-Flagship funds. FuturICT promised to develop a new interdisciplinary science of society for managing grand, interconnected, societal challenges, like financial crises, conflicts and crime, energy consumption, waves of migration, and biological and social contagion. The promised new science was expected to invest the resources of a large, federated scientific consortium and spur a major technical effort to integrate social mining and simulation modeling—i.e., the what-is description of the world state in interconnected domains with the what-next anticipation of future trajectories departing from it, and the what-if evaluation of the effects of intervention destined to manage criticalities. The FuturICT proponents applied for European funds: its team rallied around the common ambition to promote collective intelligence through the creation and use of public, transparent, non-commercial, and policy-oriented instruments of understanding, anticipation, and management of criticalities.

In the first step of the evaluation, FuturICT got the best scores. But at the end of the race, the European officers did not believe in the project strongly enough to follow the example of their predecessors in the mid-seventies: to invest in a grand scientific endeavor, beyond and relatively independent of the effective achievement of promised objectives. If they had, they probably would have made Europe a propulsive centre for culture and science, as it was in the seventies.

The reason for the final decision is of some interest in the present argument: it was feared that the FuturICT platforms could fall into enemy (e.g., mafiosi or jihadist) hands. The European scientific funding agency decided not to invest in the development of a vital instrument in order to prevent its use by competitors—a foolproof strategy, indeed.

1.4 Next Future

Mafiosi and jihadists have not yet (or are we sure?) provided themselves with instruments and platforms for the study of social criticalities, but many universities and research centers in North America, Japan, South Korea, and other places have. In most cases, these have been developed at laboratories of social mining and Institutes for Quantitative Social Science—for example, at the IQSS of Gary King at Harvard University.

The intensive use of social mining, in particular of text mining and of the so-called sentiment analysis, is often practiced for commercial reasons. Startups created to give clients information on the performance of securities and bonds, or on concurrent demands of goods and services, pop up every now and then. A good example is providing advice to economic entities willing to purchase actions of service or resource providers. There are startups that provide highly reliable one-month-ahead forecasting (see chapter 11). A commendable result. And a sufficient time interval for good financial speculations….

One could ask of course what could be the result of a massive use of this type of advice. As Brian Arthur observed, the fragmentary, atomistic use of anticipatory techniques can produce toxic effects and chaotic behaviors. Hence, the path to panic and crash is short—a result quite far from Trichet's purpose, and clearly the opposite of FuturICT's mission. This is a sort of crowd-pollution, rather than the realization of collective intelligence.

It is hard to say which will win in the long run: quantitative CSS—based on physical statistical models and focused on a short-term, local use of forecasting—or modeling CSS—focused on the explanation of social phenomena, and not just on their anticipation? It is unlikely that social mining on its own will help us to answer this question definitively.

References

Conte, R., & Paolucci, M. (2014). On agent-based modeling and computational social science. *Frontiers in Psychology, 5,* 668.

Conte, R. et al. (2015). When math helps (not) the social sciences. SSRN…. (Submitted to).

Croce, B. (1917/2007) Teoria e storia della storiografia. E. Massimilla & T. Tagliaferri (Eds.), *Bibliopolis*, Series: B (Vol 2). Croce: National publishing. ISBN 8870884465.

Dilthey, W. (1924). *Gesammelte Schriften.* Leipzig: Verlag von Teubner. http://gallica.bnf.fr/ark:/12148/bpt6k69442d.

Francesca, G., Mario, P., Diana F. A., & Rosaria, C. (2014). Group size and gossip strategies: An ABM tool for investigating reputation-based cooperation. *Multi-Agent-Based Simulation XV, 9002,* 104–118.

Schleiermacher, F. (1800/1868) *Monologen. Berlin 1868* (S. 21). (First edition: Berlin 1800. Second, 1810; third, 1821).

Part I
New Theories

Chapter 2
Cognitively Rich Architectures for Agent-Based Models of Social Behaviors and Dynamics: A Multi-Scale Perspective

Marco Campennì

2.1 Introduction

The field of modeling social behaviors and dynamics has a long and established tradition (from Trivers 1971; Axelrod and Hamilton 1981; to Sigmund et al. 2002; Hoffman et al. 2015). In this tradition, mathematical and analytical modeling approaches have played a major role since the field was established in early 1980s (Axelrod and Hamilton 1981), and they still play a central role at some of the best international research institutions (e.g., Prof. M. Nowak at Program for Evolutionary Dynamics, Harvard University; Prof. K. Sigmund at Faculty for Mathematics, University of Vienna; Prof. R. Boyd at School of Human Evolution and Social Change, ASU; Prof. J. Henrich at Department of Psychology and Vancouver School of Economics, University of British Columbia).

Starting from modeling simple (social) behaviors of human and nonhuman animals (e.g., "boids" flocking model, Reynolds 1987; cooperation, Axelrod 1984, primate fission-fusion dynamics, Boekhorst and Hogeweg 1994a, b; primate female dominance, Hemelrijk 1996), a new method and scientific approach to model social behaviors and dynamics has gained more and more attention and interest over the last decades, namely, agent-based modeling (ABM).

This approach (and more broadly speaking, this class of modeling techniques and tools) has proven to be very interesting and useful in many different applications.

ABM allows to deal with the heterogeneous individual units (i.e., agents) and emergent properties and dynamics.

The traditional analytical top-down perspective suggests modeling social dynamics at the population level, trying to individuate a possible equilibrium (i.e., a so-called steady state). Agent-based modeling, on the other hand, adopts the opposite perspective, i.e., the so-called bottom-up perspective, where the main effort of

M. Campennì (✉)
Arizona State University, 1711 S Rural Rd., Tempe, AZ 85281, USA
e-mail: mcampenni@gmail.com

F. Cecconi (ed.), *New Frontiers in the Study of Social Phenomena*,
DOI 10.1007/978-3-319-23938-5_2

modeler is to design and develop properties and behaviors of agents and rules governing the whole system and environmental conditions (the "environment" being a physical or a social environment, or a simple idealized space where interactions may take place) to make the system behaviors and dynamics emerge at the global (or collective) level starting from the local/individual interactions (i.e., the micro–macro relationship: see Alexander and Giesen 1987).

Some of those models have shown that very simple and local rules facilitate the emergence of complex behaviors at the collective level. This is the case with the famous flocking model from Reynolds (1987). In this model the simple definition of three local rules—namely, (1) separation (i.e., steer to avoid crowding local flock-mates), (2) alignment (i.e., steer toward the average heading of local flock-mates), and (3) cohesion (i.e., steer to move toward the average position of local flock-mates) applied to each individual within a group of agents—allows the flocking/schooling collective behavior to emerge at the group level.

These three simple rules combined with a small set of individual properties, such as the perceptive ability to calculate the distance from another individual and the individual direction of moving, may produce a complex and fascinating behavior common in different social species in the animal realm. In this way, the flocking behavior of birds, the schooling behavior of fishes, and many other social behaviors of living organisms may be explained as the result of simple local interactions.

2.2 Agent-Based Modeling

A simulated world may be used for exploring adaptation and evolutionary processes. The use of agent-based models allows us to improve our understanding of the behavior of individuals and populations in social and evolutionary settings.

Our claim is to suggest the use of agent-based modeling as a general theoretical and methodological tool for analyzing, studying, and modeling social behaviors and dynamics in living organisms.

Agent-based modeling (ABM) is a style of computational modeling that focuses on modeling individuals, components of individuals, or heterogeneous parts of a complex system.

There are many resources available for those interested in developing or using ABM (for a list of available tools see https://www.openabm.org/page/modeling-platforms) and there are several fields of research where researchers have adopted this approach: social sciences and human behavior (Bonabeau 2002; Gilbert and Troitzsch 2005; Gilbert 2008; Epstein and Axtell 1996), ecology (DeAngelis et al. 1991), biology (Kreft et al. 1998; Campennì and Schino 2014), and animal behavior (Hemelrijk 2000; Bryson et al. 2007).

Agent-based models are simulations based on the global consequences (macro-level) of local interactions of members of a population (micro-level). These agents (or individuals) might represent plants and animals in an ecosystem, vehicles in traffic, orpeople in crowds.

Typically, ABMs consist of an environment or framework in which individuals interact and are defined in terms of their behaviors (by procedural rules) and characteristic parameters (i.e., individual properties).

In such models, the characteristics of each individual are monitored over the time; this differs from other modeling techniques where the characteristics of the population are "averaged" and the model attempts to simulate changes in these averaged characteristics at the whole population level.

Some agent-based models are also spatially explicit: this means that individuals are associated with a location (i.e., in a geometric space). Some spatially explicit individual-based models (which is an alternative way to refer to agent-based models, often preferred in ecological and biological scientific domains) also exhibit mobility, where individuals can move around, e.g., exploring the environment or looking for sources of food.

There are three main benefits of ABM over other modeling:

- ABM captures emergent phenomena;
- ABM provides a natural description of a system;
- ABM is flexible.

Emergent phenomena result from interactions of individuals. They cannot simply be reduced to the system's parts; the whole, in this case, is more than the sum of its single parts, and this is possible because the parts interact in a complex way.

A phenomenon that emerges can have its properties' values modified in a nonlinear way; this crucial factor makes emergent phenomena very difficult to understand and predict (e.g., they can be counterintuitive).

In ABM, the researcher models and simulates the behavior of the system's constituent units, namely, agents, and their interactions and behaviors, capturing emergence from the bottom-up.

ABM is implemented as a software: the formulation, design, and implementation of algorithms, procedures and data structures needed to run an ABM force the researcher to describe the natural phenomenon or system in a very natural way.

This description is also in itself new theory generation: as in other scientific domains theory formulation is made possible by means of natural language sentences or mathematical formula; in ABM, the programming language code itself "is" the new theory.

ABM is flexible in different ways. This means that the same model can be used to investigate different aspects of the same real phenomenon or system (e.g., by modifying some model parameters); but this also means that different ABMs can be used in investigating the same topic from different perspectives to explore its multiple dimensions (e.g., evolutionary, behavioral, or cognitive).

2.2.1 Social Behavior and Communication in Living Organisms

Social behavior and cognition in living organisms are characterized by a certain number of different abilities, such as social learning, gaze following, and imitation; moreover, some living organisms exhibit a complex communication system which allows them to express a wide range of emotions, moods, social relationships, and mental representations.

Human language can be considered as a tangled web of syntax, semantics, phonology, and pragmatic processes. All of these components work together, allowing language to emerge; we can find most of them (perhaps in different forms) in other animals. We can make a rough classification of these mechanisms, identifying three different classes of processes: (1) signaling, (2) semantics, and (3) syntax.

Signaling includes all of perceptual and motor systems underlying speech and signing; semantics may be considered as the central cognitive mechanism that supports the formulation of concepts and their expression and interpretation; syntax represents the mechanism that allows animals to generate structures and to map between signals and concepts.

Signals and semantics have strong social components: the former are used in communication and must be learned and shared among community member and require sophisticated abilities in order to imitate complex signals; the latter require the ability to infer the signaler's intentions by more-or-less indirect cues.

Scientific research in comparative cognition aims at studying different species to reveal similarities and differences in each cognitive mechanism; the investigation includes the study at multiple levels of description, from the genetic to neural and then behavioral level. Hypotheses about the evolution of cognition can be generated and tested from found similarities, both in terms of homology and analogy.

Only recently, researchers working in the field of comparative social cognition have started to considerate non-primate mammals (e.g., dogs, rats, goats), many bird species (among corvids, jays, crows, ravens), reptiles, fish, and social insects to investigate cognitive abilities and skills needed for social interactions (for a detailed table of taxonomic information, see Table 2.1 [reproduced from Fitch et al. 2010]).

Results obtained with these species often revealed surprising cognitive abilities: dogs or ravens succeeded in tasks when our closer non-human primate relatives failed. These kind of results have to be taken with a grain of salt, as they reflect a view of evolutionary mechanisms in which cognitive capacities increase with a species' relatedness to humans (Striedter 2004). More modern Darwinian viewpoints postulate that a species' cognitive ability evolves to fit its cognitive niche. So we expect that the evolution of specific cognitive capacities derives from the physical and social environment: species living in environments where they have to perform complex navigation tasks will evolve sophisticated spatial memory, whereas species living in complex social communities will exhibit superior social cognition.

This perspective allows us to surmise that a convergent evolution of analogous cognitive mechanisms (analogs) will be detected in widely separated species that face similar cognitive problems.

Tab. 2.1 Species and Clades Studied in Contemporary Social Cognition Research

	Common Name	Genus	Species	Major Clade	Minor Clade
Vertebrates	Common Marmoset	*Callithrix*	*jacchus*	class Mammalia	order Primates
	Chimpanzee	*Pan*	*troglodytes*	" "	" "
	Orangutan	*Pongo*	*pygmaeus*	" "	" "
	Capuchin	*Cebus*	*apella*	" "	" "
	Rhesus Macaque	*Macaca*	*mulatta*	" "	" "
	Bottlenose Dolphins	*Tursiops*	*truncatus*	" "	order Cetacea
	Humpback Whale	*Megaptera*	*novaeangliae*	" "	" "
	Harbor Seal	*Phoca*	*vitulina*	" "	suborder Pinnipedia
	S. African Fur seal	*Arctocephalus*	*pusillus*	" "	" "
	Domestic Dog	*Canis*	*familiaris*	" "	order Carnivora
	Domestic Goat	*Capra*	*hircus*	" "	order Artiodactyla
	Greater Sac-Winged Bat	*Saccopteryx*	*bilineata*	" "	order Chiroptera
	Japanese Quail	*Cotumix*	*japonica*	class Aves	order Galliformes
	Pigeon	*Columba*	*livia*	" "	order Columbiformes
	Bald Ibis	*Geronticus*	*eremita*	" "	order Threskiornithidae
	Budgerigar	*Melopsittacus*	*undulatus*	" "	order Psittaciformes
	Kea	*Nestor*	*notabilis*	" "	" "
	African Gray Parrot	*Psittacus*	*erithacus*	" "	" "
	European Starling	*Sturnus*	*vulgaris*	" "	order Passeriformes
	Woodpecker Finch	*Cactospiza*	*pallida*	" "	" "
	Swamp Sparrow	*Melospiza*	*georgiana*	" "	" "
	Zebra Finch	*Taeniopygia*	*guttata*	" "	" "
	Bengalese Finch	*Lonchura*	*striata domestica*	" "	" "
	New Caledonian Crow	*Corvus*	*moneduloides*	" "	family Corvidae
	Raven	*Corvus*	*corax*	" "	" "
	Rook	*Corvus*	*frugilegus*	" "	" "
	Scrub Jay	*Aphelocoma*	*californica*	" "	" "
	Archerfish	*Toxotes*	*chatareus*	infraclass Teleostei	family Toxotidae
	Red-footed Tortoise	*Geochelone*	*carbonaria*	class Reptilia	family Testudinae
Nonvertebrates	Octopus	*Octopus*	*vulgaris*	phylum Mollusca	class Cephalopoda
	Honeybee	*Apis*	*mellifera*	class Insecta	order Hymenoptera

This table provides taxonomic information regarding the species discussed in this review. Only the common name is used in the main text. The major and minor clades help to contextualize the phylogenetic position of these species utilizing traditional Linnaean classification, even when (as for class "Reptilia") this traditional grouping is polyphyletic.

2.2.2 Communication, Social Cognition and Theory of Mind (ToM)

Can non-human animals have a theory of mind? The debate is still open, but since Premack and Woodruff asked, "Does the chimpanzee have a theory of mind?" in their seminal paper (Premack and Woodruff 1978), the interest of researchers has steadily increased (Povinelli and Vonk 2003; Tomasello et al. 2003).

Even if some earlier results obtained testing the cooperative behavior of primates in tasks where they must trustingly interact with human experimenters showed little evidence of ToM in chimpanzees (Povinelli and Eddy 1996; Povinelli et al. 1990), more recent competitive experiments showed unexpected strong results (Hare et al. 2000; Hare 2001). In these experiments subjects competed with other conspecifics and/or human experimenters for sources of food and results probably derive from the more ecological significance of the task for primates.

A large amount of data obtained from experiments using a wide range of different primates (Braeuer et al. 2007; Karin-D'Arcy and Povinelli 2002; Kaminski et al. 2008) suggests that in most cases primates can distinguish between conspecifics who know where some sources of food are hidden from "guessers," who know that food has been hidden, without knowing exactly where.

Corvids tested with similar tasks (Clayton et al. 2007) showed a strong use of sophisticated cognitive mechanisms. Both scrub jays and ravens can differentiate between competitors that have or have not seen food cached in particular locations, modifying their strategy or behavior in accordance with information retrieved using ToM (Bugnyar and Heinrich 2005, 2006; Emery and Clayton 2001; Dally et al. 2005, 2006).

We can assume that some primates and corvids can consider perceptions of others in using information derived by interaction with them and the environment to infer possible consequences of others' actions in food-related tasks.

Finally, some results seem to suggest that chimpanzees and corvids are capable of attributing certain mental states to others (Call and Tomasello 2008; Clayton et al. 2007), even if they are not able to deal with false beliefs like humans do. In this sense, scientific studies of avian cognition (and not only the study of primate cognitive abilities) can help us to better understand the evolution of advanced socio-cognitive skills.

Nevertheless, there are many different elements contributing to the success of such kinds of tasks; cooperative behaviors and complex interactions between individuals can emerge from simple individual aptitudes or motivations. So it is not clear at all wherein and when cognitive abilities (such as ToM) are strictly necessary to solve these kind of tasks; it may be sufficient to integrate perceived information with some simple heuristics to solve quite complex food-related tasks. Moreover, experience (both in terms of past interaction with others and familiarity with a specific task) plays a very important role in developing social intelligence.

2.2.3 "Animal Culture" and Imitation

Evolutionary biologists study the evolution of cultural artifacts, related cognitive abilities, and processes because these kinds of phenomena represent a very good example of a system's operating by inheritance and adaptation. Moreover, cultural transmission processes are more rapid than genetic ones, and the study of "culture" in animals can allow us to better understand and identify evolutionary roots of cultural processes in humans, possibly the most cultural animals on the earth.

Cultural evolution works in a way that is very similar to biogenetic evolution (Mesoudi et al. 2004), following some principles and dynamics already identified by Darwin (Darwin 1964) more than 150 years ago. In this context, language is a very good example of this kind of historical change (Fitch 2008), and linguistic elements (words and grammatical rules) can be studied and analyzed using tools and instruments borrowed from molecular phylogenetics (Cavalli-Sforza et al. 1992; Lieberman et al. 2007; Pagel et al. 2007). A very distinctive mechanism of cultural phenomena is their cumulative nature: ideas, especially good ideas, can be accumulated within the same generation and transmitted to the next, following a principle of high-accuracy copying very similar to that adopted to explain genetic transmission. Accumulation of high-fidelity elements in animal species is a topic still open to debate in the study of cumulative change and evolution of culture (Heyes 2009; Huber et al. 2009; Tennie et al. 2009).

The relation between culture and social learning could be very interesting and stimulating for researchers studying social behavior in animals. Some results suggest that social learning is possible in group-living mammals (Heyes 1994), birds (Zentall 2004), fish (Schuster et al. 2006), and insects (Leadbeater and Chittka 2007); however, we don't have sufficient information about the evolutionary roots of these abilities, and even if some eminent researchers have hypothesized about the social origin of intelligence (see Dunbar and Shultz 2007), in some cases non-social species have also shown the same ability to learn to solve a task by observing actions performed by a conspecific (see Wilkinson et al. 2010, where solitary tortoises can solve a detour task after the observation of a conspecific completing the task).

In this view, imitation can be viewed as the non-genetic reply to the inheritance of phenotypic attributes in supporting cultural phenomena. However, it is less clear what types of imitation can play this role in cumulative culture. Surely, imitation has to be as accurate as possible in the copying process and it must involve certain forms of learning, i.e., the ability to acquire new skills and behaviors.

Moreover, observation of someone else's behavior has to be selective, as shown by theoretical models of adaptive advantages of social learning (Galef and Laland 2005). An individual who observes the behavior of others has to consider the specific relationship existing between the target individual and her- or himself (i.e., dominance, affiliation, tolerance) in order to perform the correct action; thus, the ability to correctly monitor the behavior of others is a crucial element of any social behavior (cooperation, communication, and competition). Environmental and physical conditions may limit the individual's capacity to observe every animal and actions performed within a specific social group; for this reason, selectivity is also very crucial for acquisition and spreading of social information.

2.2.4 Information Exchange

Information is the vital component for the emergence of communication and communicative systems. It may be transmitted, processed, and used to make decisions and to coordinate actions or individuals.

The transmission of information may be related to the existence of a system that allows an individual to signal something to someone else: in this case, emitted signals have to be exchanged in a coordinated way, preserving the original content. Nevertheless, the transmission of information may also occur in an unintentional way: the individual behavior of performing a specific task (e.g., searching for food in a particular place) can be used as a behavioral cue by other observing individuals. In nature, we can find a wide range of possible signaling systems that have evolved over the time to permit the exchange of information at very different levels, from very micro entities to macro ones: e.g., from quorum signaling in bacteria (Schauder and Bassler 2001; Taga and Bassler 2003; Kaiser 2004) through the dance of the honeybees (Dyer and Seeley 1991), birdcalls (Hailman et al. 1985; Evans et al. 1993; Charrier and Sturdy 2005) and alarm calls in many different species (Cheney and Seyfarth 1990; Seyfarth and Cheney 1990; Green and Maegner 1998; Manser 2002) and, finally, to human language (Fitch 2010; Cangelosi 2001). The emergence of communicative systems facilitates the evolution of social structures and dynamics in animals.

2.2.5 Agent-Based Modeling of the Evolution of Communicative Systems

Some researchers have proposed to study the evolution of signaling systems as sender–receiver games (Skyrms 2009), stressing the fact that such games are simple, tractable models of information transmission and that they provide a basic setting for studying the evolution of meaning. In these models it is easy to investigate not only the equilibrium structure, but also the dynamics of evolution and learning.

Some previous studies of the adaptive nature of communication for coordination found communication beneficial; others, not. Schermerhorn and Scheutz (2007) claim that this results from the lack of a systematic examination of important variables such as (i) communication range, (ii) sensory range, and (iii) environmental conditions. These authors presented an extensive series of simulative experiments where they explored how these parameters affect the utility of communication for coordination in a multi-agent territory-exploration task.

A very useful review of recent progress in computational studies investigating the emergence of communication among agents via learning or evolutionary mechanisms was published by Wagner et al. (2003). In this work, Wagner and colleagues presented a review of issues related to animal communication and the origins and evolution of language. The studies reviewed show how different elements (as population size, spatial constraints on agent interactions, and the specific tasks agents have to face) can all influence the nature of the communication systems and the ease with which they are learned and/or evolve. The authors identify some important areas for future research in the evolution of language, including the need for further computational investigation of key aspects of language such as open vocabulary and the more complex aspects of syntax.

Alarm-calling behavior in animals is one of the most intriguing behaviors exhibited by a wide range of animals, and the study of such behavior may allow us to better understand the evolutionary roots of human language. Noble and colleagues (Noble et al. 2010) proposed a model of alarm-calling behavior in putty-nosed monkeys, stressing the need for real data to determine whether a computational model is a good model of a real phenomenon (or behavior). They argued that computational modeling, and in particular the use of agent-based models, is an effective way to reduce the number possible explanations when competing theories exist. According to their approach, simulations may achieve this both by classifying evolutionary trajectories as either plausible or implausible and by putting lower bounds on the cognitive complexity required to perform particular behaviors. Of course, this last point has a lot of implications for many fields of investigation (e.g., the study of bounded rationality). The authors use the case-study method to understand whether the alarm calls of putty-nosed monkeys could be a good model for human language evolution.

In a previous article (Noble 1999), one of the same authors presents a general model that covers signaling with and without conflicts of interest between signalers and receivers. In this work, simple game-theoretic and evolutionary simulation models are used to suggest that signaling will evolve only if it is in the interests of both parties.

As we made clear above (see section 2.2.3 about animal culture), another critical issue concerns the relationship between gene and culture co-evolution. It has been argued that aspects of human language are both genetically and culturally transmitted. Nevertheless, how these processes might interact to determine the structure of language is not very clear yet. Agent-based modeling can be used to study gene-culture interactions in the evolution of communication. Smith (2002) presented a model showing that cultural selection resulting from learner biases can be crucial in determining the structure of communication systems transmitted through both genetic and cultural processes. Moreover, the learning bias that leads to the emergence of optimal communication systems in the model resembles the learning bias brought to the task of language acquisition by human infants. This result seems to suggest that the iterated application of such human-learning biases may explain much of the structure of human language.

Finally, a well-constructed presentation of different types of models implemented to study the evolution of communication and language was made in Cangelosi (2001). In this study, the distinction among signals, symbols, and words is used to analyze evolutionary models of language. In particular, the work shows how evolutionary computation techniques, such as the Artificial Life approach (artificial neural networks and evolutionary algorithms), can be used to study the emergence of syntax and symbols from simple communication signals. First of all, the author presents a computational model that evolves repertoires of isolated signals. In the model presented, the case study is the simulation of the emergence of signals for naming foods (good and bad sources of food) in a population of foragers. Then, another model is implemented to study communication systems based on simple signal–object associations. Finally, models designed to study the emergence of grounded symbols are discussed in general, including a detailed description of a work on the evolution of simple syntactic rules. In the paper, several important issues (such as symbol–symbol relationships in evolved languages and syntax acquisition and evolution) are discussed, and computational models are used to suggest an operational definition of the signal/symbol/word distinction and to better understand the role of symbols and symbol acquisition in the origin of language.

2.2.6 Agent-Based Modeling of Social Organization, Structures, and Dynamics in Living Organisms

One of the most important aspects of all biological systems is the ability to cooperate. Complex cooperative interactions are required for many levels of biological organization, ranging from single cells to groups of animals (Hamilton 1964; Trivers 1971; Axelrod and Hamilton 1981; Wilson 1975).

How can natural selection lead to cooperation? This kind of question has fascinated evolutionary biologists since Darwin (Darwin 1964; Trivers 1971; Hammerstein 2003). Cooperation among relatives is usually explained by adopting the concept of kin selection: it represents the idea that selfish genes lead to unselfish phenotypes (Frank 1989; Hamilton 1963).

Concerning the evolution of cooperation among genetically unrelated individuals, various mechanisms have been proposed based on (evolutionary) game theory (Doebeli and Hauert 2005): cooperators form groups and thus they preferentially interact with other cooperators (Sober and Wilson 1998; Wilson and Sober 1994); cooperators occupy spatial positions in topological structures (e.g., lattices or networks) and interact with their neighbors—who are also cooperators (Hauert 2001; Killingback et al. 1999; Nowak and May 1992); reputation may facilitate the evolution of cooperation via indirect reciprocity (Alexander 1987; Nowak and Sigmund 1998) or punishment (Sigmund et al. 2001).

From insects to animals, the social behavior shows complex relationships between individuals and interesting effects at the population level of very local interactions.

Eusociality, i.e., the phenomenon by means of which some individuals reduce their own lifetime reproductive potential to raise the offspring of others, underlies the most advanced forms of social organization and the ecologically dominant role of social groups of individuals (from insects to humans). For more than 40 years kin selection theory, based on the concept of inclusive fitness (in evolutionary biology and evolutionary psychology, inclusive fitness is the sum of an organism's classical fitness—how many of its own offspring it produces and supports—and the number of equivalents of its own offspring it can add to the population by supporting others), has been the major theoretical explanation for the evolution of eusociality.

Nowak and colleagues (2010) showed the limitations of this approach, arguing that standard natural selection theory in the context of precise models of population structure could represent a simpler and better approach. This new perspective allowed the evaluation of multiple competing hypotheses and provided an exact framework for interpreting empirical observations.

In the animal kingdom, a well-known form of cooperative/altruistic behavior may be found in the social organization of vampire bats—more precisely, the blood-sharing activity among vampire bats.

In this pro-social behavior, of particular interest is the specific formation and maintenance of (new) social structures (i.e., roosts) from initial populations as a consequence of both (i) demographic growth and (ii) social organization. This specific example is especially interesting because of the flexible nature of roost-switching behavior shown by these animals in natural wild conditions.

A very interesting agent-based model of such natural phenomenon is described in Paolucci et al. (2006). In this work, the main hypothesis concerns the role of grooming networks in roost formation, and the investigations are performed by means of agent-based simulations based on ethological evidence (i.e., using real data to parametrize the model).

The use of simulation allows the authors to discuss generative hypotheses concerning the origin of roosts, which can emerge from individual behavior. Results obtained not only confirm the main expectations but also reveal the need for a natural ordering in grooming-partner selection. This specific ordering can be obtained not only through (i) kin-based groups but also through (ii) the maintenance of a non-kin–based precedence rule.

Individuals of most social species (even guppies) keep track of how their groupmates have treated them in the past, but only some of these social species are able to exhibit complex social behavior, complex relationships, and dynamics between individuals.

Primates, for instance, appear to also keep track of how their troop-mates treat each other. This takes much more memory, and possibly compositional reasoning; generally speaking, it requires more sophisticated cognitive abilities.

Many researchers have proposed agent-based models of social behavior and organization in different species. Several publications concern the social behavior and dynamics of non-human primates, both for the intrinsic complex nature of social behaviors in primates and for a wide range of similarities between human and non-human primates activities. Hemelrijk and Bryson (see Hemelrijk 2000; Bryson et al. 2007) presented very interesting agent-based models of social organization in non-human primates based on dominance-ranking dynamics and relationships and gender differences (e.g., in terms of aggressive behavior propensity).

2.3 Why Do We Need Cognitive Agents?

Group-living animals often exhibit complex (and/or complicated) social behaviors. Sometimes the pure observation of such behaviors is not sufficient to improve our understanding of reality and the best option may be to use modeling to capture essential elements of the real phenomenon or system and to better understand the dynamics governing the whole process (as shown in Campennì and Schino 2014—see also Figs. 2.1 and 2.2), the theoretical assumption about a possible simple cognitive mechanism of memory governing the reciprocal exchange of cooperative behavior and the mechanism of partner choice can represent an useful way to understand the origins of cooperation in living organisms both from a behavioral and an evolutionary perspective. The hypotheses that non-cognitive mechanisms and dynamics may allow the cooperation to emerge and spread in populations of group living animals has been shown to represent a valid alternative from the behavioral point of view (i.e., they can deal with cooperation “hic and nunc”), but not from the evolutionary perspective (i.e., they are not able to explain the related evolutionary process).

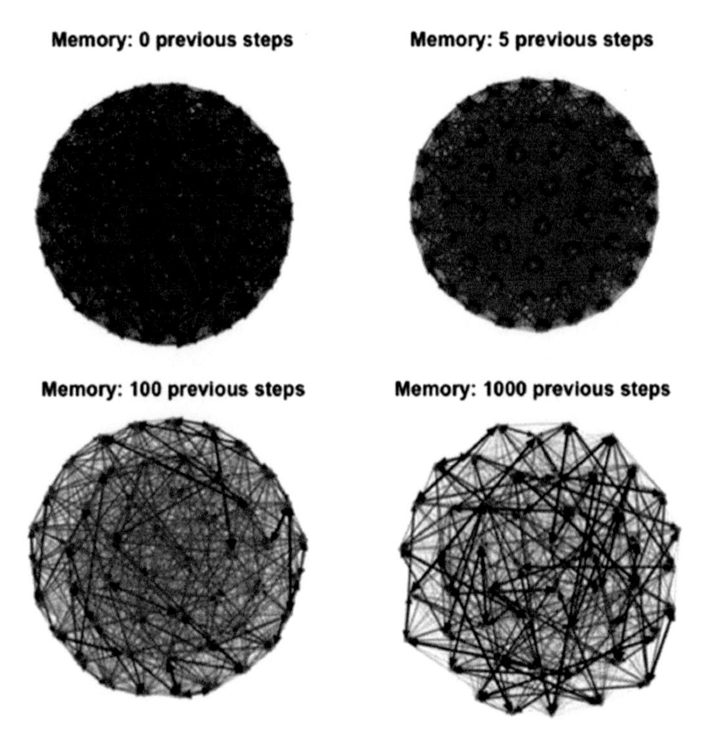

Fig. 2.1 Social network analysis of sociometric matrices obtained as output of an ABM of cooperative behavior based on partner choice

2.3.1 Cooperation Theory and ABM

The exchange of cooperative behaviors is a common feature of different animal societies. This is particularly true for those species that form stable social groups, where exchanges of cooperative behaviors (e.g., grooming, food tolerance or aggressive coalitions) are frequently observed (Dugatkin 1997; Cheney 2011).

The analysis of how group-living animals distribute their cooperative behaviors among group mates has revealed some common features that can be observed across a variety of settings and species.

First, group-living animals show differentiated social relationships: each group member interacts/cooperates frequently with some group mates and rarely, if ever, with others.

As a result, pairs of animals from the same social group may widely differ in their frequency of interaction.

Second, a positive relation is often found across pairs between cooperation given and received (Schino 2007; Schino and Aureli 2008; Seyfarth and Cheney 2012).

Among the several hypotheses that theoretical and evolutionary biologists have proposed to explain the evolution of cooperative behaviors (West et al. 2007) reciprocity is perhaps the most debated, and reviews of its empirical evidence have

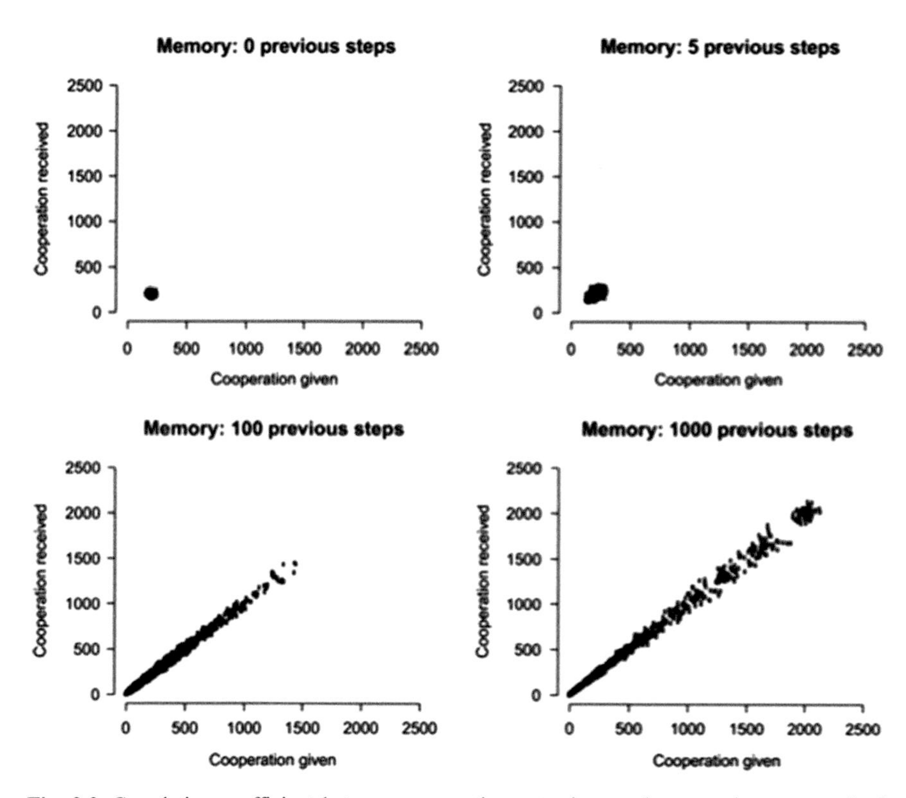

Fig. 2.2 Correlation coefficient between cooperative acts given and cooperative acts received calculated on sociometric matrices obtained as output of an ABM of cooperative behavior based on partner choice

reached widely diverging conclusions (Cheney 2011; Clutton-Brock 2009; Schino and Aureli 2009).

Part of this confusion stems from a failure to appreciate that two different processes can underlie reciprocation.

The first process, that we could define as "temporal relations between events", is a strictly within-pair process: individual A behaves cooperatively towards individual B in relation to how B has previously behaved towards A. Each A–B dyad is conceptually isolated from all others (i.e., the presence and behavior of other individuals do not affect the behavior of A–B).

This is essentially equivalent to Bull and Rice (1991) "partner-fidelity model", to Noë (2006) "partner control model" and to classical reciprocal altruism (Trivers 1971). The classical iterated prisoner's dilemma belongs to this category of models.

The second process, that we could define as "partner choice based on benefits received", is an across-pair process with a strong comparative component: individual A behaves cooperatively towards individual B rather than individual C in relation to a comparison of how B and C have behaved towards A.

This is essentially partner choice based on outbidding competition (Noë and Hammerstein 1994) and it is equivalent to Bull and Rice (1991) and to Noë (2006) "partner-choice model" (see also Eshel and Cavalli-Sforza 1982 for an earlier study).

Empirical evidence shows reciprocal exchanges of cooperative behaviors depend more commonly on partner choice based on benefits received than on within-pair temporal relations between events (Tiddi et al. 2011; Fruteau et al. 2011).

Despite its prevalence, partner choice has been widely neglected as a general explanation for the evolution of cooperative behaviors (Sachs et al. 2004).

Theoretical modeling has focused mostly on the analysis of within-pair temporal relations between events, and a vast literature exists on the possible strategies that can promote the evolution of cooperation through this process (Bshary and Bronstein 2011; Nowak 2006; Nunn and Lewis 2001; Lehmann and Keller 2006; André and Baumard 2011).

In contrast, theoretical models of the evolution of cooperation by partner choice are comparatively rare. In some of the few existing examples, partner choice is based on the general tendency of potential partners to cooperate, rather than on actual cooperation received by each partner (Barclay 2011; Roberts 1998). As such, these models seem more relevant to indirect than to direct reciprocity. In other modeling attempts, partner choice is included in the form of the possibility to terminate a within-pair series of cooperative interactions (Sherratt and Roberts 1998; Johnstone and Bshary 2002). The relative lack of models of partner choice based on benefits received is puzzling, considering its obvious relevance for group living animals.

When developing a theoretical model of a biological phenomenon or system, one can aim either at reproducing important features of the target system "as is" (for cooperative exchanges, see Roberts 1998), or at modeling its evolution, i.e., at reproducing the changes that would occur across generations as a result of natural selection (Axelrod and Hamilton 1981).

Ideally, however, a good model should be able to reproduce both aspects of the phenomenon and if (and only if) both tests are successful a stronger case for the relevance of the principles underlying the model in explaining the target system being modeled could be made.

In this perspective ABM represent the ideal modeling tool candidate to accomplish both tasks.

2.3.2 Why Agent-based Models and not Other Modeling Approaches?

2.3.2.1 Bottom-up Vs top-down Approach

We already tried to stress this point above in this chpater. The main difference between traditional analytical modeling (AM) and ABM is that the former is a perfect example of top-down approach, while the latter is based on the bottom-up "philosophy".

AMs simplify the real phenomenon or system as much as possible to identify the minimal requirements (e.g., parameters) allowing the model to exhibit (almost)

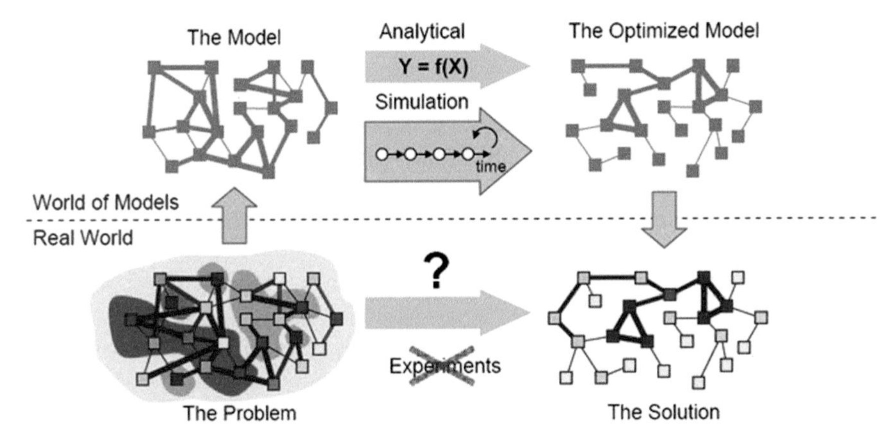

Fig. 2.3 A comparison of analytical and simulative approaches

the same behavior (i.e., dynamics and equilibria) we can observe in reality. This of course implies that strong assumptions and simplifications need to be made in order to make the model manageable from an analytical point of view (i.e., the model needs to be solvable—there should be a solution). So, for instance, cooperation can be modeled as one of different possible outputs of a two players game, as in the case of the Prisoner Dilemma (Axelrod 1984).

ABM adopts the opposite approach, where in principle there is no limit in the definition of heterogeneous properties of individual units or agents; each agent can be different, can behave differently and can interact with other agents and the environment in a different way. The emergent behavior of the whole system is possible because the definition of individual properties, behaviors and relationships (i.e., the micro-macro path).

Moreover, ABM allows us to deal with complex systems and dynamics that per definition exhibit a non-linear behavior. In this perspective, ABM represents a useful approach to model feedbacks, loops and complex causal relationships (e.g., the downward causation Kim 1992) (Fig. 2.3)

2.3.2.2 Too-Complex Dynamics and Behaviors (e.g., Impossible To Model Using an Analytical Approach)

Sometimes ABM can be the only way to model a real phenomenon or system. ABM allows us to deal with a limited number of variables and parameters, simply because the analytical model itself would not be solvable otherwise; moreover, if there is no single solution to the analytical model, ABM is often used to approximate the behavior of the original model by simulating a certain amount of times (i.e., runs) the behavior occurs starting from randomly selected initial conditions (i.e., combinations of parameters).

2.3.2.3 In-Silico Data Generator

ABM can be looked at as an in-silico generator of data—a huge amount of data that could hardly be collected using an empirical or observational approach.

Of course, since they are affected by initial stochastic conditions, ABMs need to be run several times using the same set of parameters then averaged; but the very good news is that such outputs can be then used as any other kind of empirical data, applying exactly the same statistical analyses and metrics. Thus, the advantage in using ABMs as generators of data is very clear.

I had the chance to explore the usefulness and scientific relevance of using ABM in investigating social behaviors and dynamics in living organisms. In the following sections of this chapter I will present additional interesting case studies to show how the adoption of the ABM approach to investigating social behaviors and dynamics (henceforth SBD) can also be relevant and useful in other scientific domains and at different scales.

2.4 Social–Ecological Systems

The study of the complex social–ecological systems (SES) was inaugurated a few decades ago by a group of ecologists and economists interested in investigating the complex interconnections and tensions between (complex) ecological systems and (complex) social systems living in, and operating on, such ecologies.

A social–ecological system consists of a combination of biological, geological, and physical units and associated social actors and institutions.

Social–ecological systems present some specific characteristics, such as complexity and adaptation, and are delimited by spatial or functional boundaries surrounding particular ecosystems (Fig. 2.4).

A social-ecological system can be defined as (Redman et al. 2004, p. 163):

- A coherent system of biophysical and social factors that regularly interact in a resilient, sustained manner;
- A system that is defined at several spatial, temporal, and organizational scales, which may be hierarchically linked;
- A set of critical resources (natural, socioeconomic, and cultural) whose flow and use is regulated by a combination of ecological and social systems; and
- A perpetually dynamic, complex system with continuous adaptation. (Machlis et al. 1997; Gunderson and Holling 2002; Berkes et al. 2003)

Researchers have used the concept of social–ecological systems to emphasize the integrated concept of humans in nature and to stress that the distinction between social systems and ecological systems is artificial and arbitrary (Berkes et al. 2001)

While resilience has a somewhat different meaning in social and ecological contexts (Adger 2000), the SES approach holds that social and ecological systems are linked through feedback and loop mechanisms, and that both display resilience and complexity (Berkes et al. 2003).

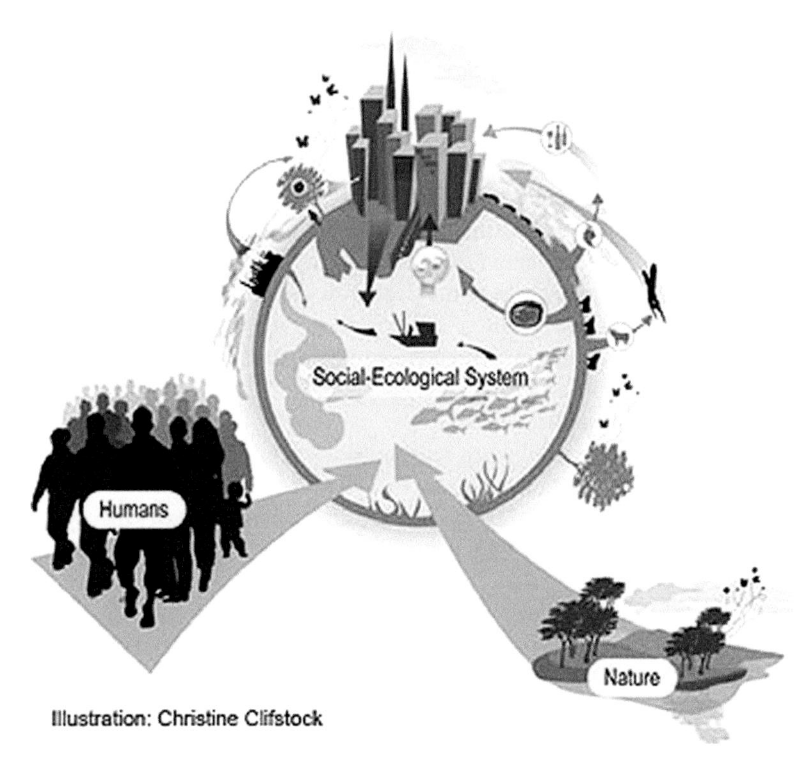

Fig. 2.4 An illustration of the concept of SES from *Sustainable Development Update,* Issue 1, Volume 8, 2008. The *Sustainable Development Update* (SDU) focuses on the links among ecology, society, and the economy. It is produced by Albaeco, an independent, nonprofit organization. *SDU* is produced with support from Sida, the Swedish International Development Cooperation Agency, Environment Policy Division (Dr. Fredrik Moberg, Editor)

Until the appearance of this new field of scientific investigation a few decades ago, the interaction and co-operation between social sciences and natural sciences was very limited and thus so was the study of social-ecological systems. While, on the one hand, traditional ecologists had tried to ignore human actions in specific ecologies, on the other hand, researchers from social science disciplines had ignored the environment's role in defining and affecting human activity (Berkes et al. 2003).

Although some eminent thinkers (e.g., Bateson 1979) had tried to fill the gap between natural and cultural spheres, the larger part of sociological studies focused on investigating processes within the social domain only, considering ecosystems as "black boxes" (Berkes et al. 2001) and assuming that if the social system performs adaptively or is well organized institutionally it will also manage the environmental resource base in a sustainable fashion (Folke 2006).

This framework changed through the 1970s and 1980s with the rise of several social sciences subfields explicitly including the environment (Berkes et al. 2003).

Among them, environmental ethics arose from the need to develop a philosophical framework for relations between humans and the environment where they live

(Berkes et al. 2001); political ecology, from the need to expand ecological concerns to the include cultural and political activity within an analysis of socially constructed ecosystems (Greenberg and Park 1994); environmental history, from the growing collection of material documenting different relationships between societies and their environment; ecological economics, from examining the link between ecology and economics and integrating the scientific framework of economics within the concept of the ecosystem (Costanza et al. 2001); common property, from the integration of resource management and social organization and the analysis of how institutions and property-rights systems deal with the dilemma of the "tragedy of the commons" (McCay and Acheson 1987; Berkes 1989); and, finally, from traditional ecological knowledge, which refers to ecological understanding built, not necessarily by experts, but more simply by people who live and use the resources of a particular place (Warren et al. 1995).

Each of the six summarized areas represents a scientific "bridge," combining a natural science and social science perspective (Berkes et al. 2003).

SES theory had been inspired by systems ecology and complexity theory; nevertheless, the concept of SES does not conceptually overlap those of system ecology and complex systems.

The studies of SES include some central societal concerns (e.g., equity and human well-being) that have traditionally received little attention in complex adaptive systems theory (It should be noted that there is growing attention in this direction and an academic proliferation of institutes and programs integrating the former with the latter—see, e.g., the Julie Ann Wrigley Global Institute of Sustainability at ASU, the Atkinson Center for a Sustainable Future at Cornell University, or the Sustainable Systems Program of the School of Natural Resources and Environment at University of Michigan); and conversely, there are areas of complexity theory (e.g., quantum physics) that have little direct relevance for understanding SES (Cumming 2011) (at least thus far to our knowledge).

SES theory incorporates ideas from theories relating to the study of resilience, sustainability, robustness, and vulnerability (e.g. Levin 1999; Berkes et al. 2003; Gunderson and Holling 2002; Norberg and Cumming 2008). While SES theory draws on a range of discipline-specific theories, such as island biogeography, optimal foraging theory, and microeconomic theory, it is much broader than any of these individual theories alone (Cumming 2011).

Because of its recent development and scientifically young age, SES theory has emerged from a combination of disciplinary platforms (Cumming 2011), while the notion of complexity developed through the work of many scholars, notably the Santa Fe Institute (2002). From this perspective it could be argued that complex system theory is one of the most important "intellectual parents" of SES (Norberg and Cumming 2008). However, due to the social context in which SES research operates, and the potential (and sometimes actual) impact of SES research on policy recommendations that will have consequences on real people's lives, SES research has been considerably more "self-sustaining" and more "pluralistic" than complexity theory has ever acknowledged (Cumming 2011).

Studying SESs from a complex-system perspective is a fast-growing interdisciplinary field which can be interpreted as an attempt to link different disciplines into a new body of knowledge that can be applied to solve some of our most serious actual environmental problems (Cumming 2011).

Management processes in the complex systems can be improved by making them adaptive and flexible and able to deal with uncertainty and volatility, and by building in the capacity to adapt to change. SESs are both complex and adaptive, meaning that they require continuous testing and study in order to develop the knowledge and understanding needed in order to cope with change and uncertainty (Carpenter and Gunderson 2001).

A complex system differs from a simple system in that it has a number of attributes that cannot be observed in simple systems, such as emergence, self-organization, non-linearity, uncertainty, and scale (Berkes et al. 2003; Norberg and Cumming 2008). As argued above, ABM is one of the best conceptual and scientific tools to implement models that are able to deal with such properties.

Emergence is the appearance of behavior that could not be anticipated from knowledge of the parts of the system alone.

Self-organization is one of the defining properties of complex systems. The basic idea is that open systems will reorganize themselves at critical points of instability. Holling's adaptive renewal cycle is an illustration of reorganization that takes place within the cycles of growth and renewal (Gunderson and Holling 2002). The self-organization principle, operationalized through feedback and loop mechanisms, applies to many biological and social systems and even to mixtures of simple chemicals. High-speed computers and nonlinear mathematical techniques help simulate self-organization by yielding complex results and yet strangely ordered effects. The direction of self-organization will depend on such things as the system's history; it is path dependent and difficult to predict (Berkes et al. 2003).

Nonlinearity is related to fundamental uncertainty (Berkes et al. 2003). It generates path dependency, which refers to local rules of interaction that change as the system evolves and develops. A consequence of path dependency is the existence of multiple basins of attraction in ecosystem development and the potential for threshold behavior and qualitative shifts in system dynamics under changing environmental influences (Levin 1998).

Scale is important when dealing with complex systems. In a complex system many subsystems can be distinguished; and since many complex systems are hierarchic, each subsystem is nested in a larger subsystem (Allen and Starr 1982). For instance, a small watershed may be considered an ecosystem, but it is a part of a larger watershed that can also be considered an ecosystem and an even larger one that encompasses all the smaller watersheds (Berkes et al. 2003). Phenomena at each level of the scale tend to have their own emergent properties, and different levels may be coupled through feedback relationships (Gunderson and Holling 2002). Therefore, complex systems should always be analyzed or managed simultaneously at different scales.

2.4.1 Role of Traditional Knowledge in SES

Berkes and colleagues (Berkes et al. 2001) distinguish four sets of elements which can be used to describe social–ecological system characteristics and linkages: ecosystems, local knowledge, people and technology, property-rights institutions (for an updated theory of role of traditional knowledge see Fig. 2.5 from Tengö, M., Brondizio, E. S., Elmqvist, T., Malmer, P., & Spierenburg, M. (2014). Connecting diverse knowledge systems for enhanced ecosystem governance: the multiple evidence base approach. Ambio, 43(5), 579–591.

Acquiring knowledge about SESs is an ongoing, dynamic learning process, and such knowledge often emerges within institutions and organizations. The effectiveness of this process requires the involvement of institutions and may be implemented by means of multi-level social networks. It is thus the communities which interact with ecosystems on a daily basis and over long periods of time that possess the most relevant knowledge of resource and ecosystem dynamics, together with associated management practices (Berkes et al. 2000). Some scholars have suggested that management and governance of SESs may benefit from a combination of different knowledge systems (McLain and Lee 1996); others have attempted to import such knowledge into the scientific knowledge field (Mackinson and Nottestad 1998). There are also those who have argued that it would be difficult to separate these knowledge systems from their institutional and cultural contexts (Berkes 1999) and those who have questioned the role of traditional and local knowledge

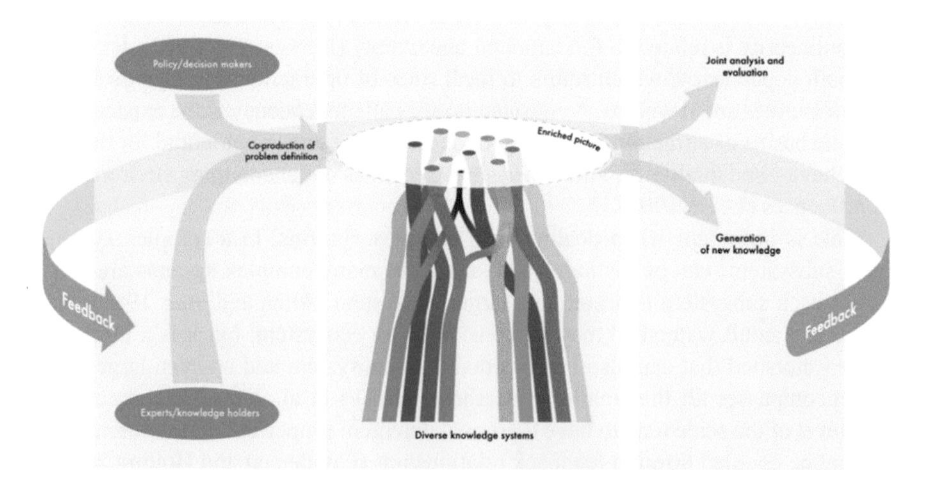

Fig. 2.5 Outlining three phases of a multiple evidence base approach, emphasizing the need for co-production of problem definitions as well as joint analysis and evaluation of the enhanced picture created in the assessment process. Phase 1 concerns defining stakeholders, problems, and goals in a collaborative manner. Phase 2 entails bringing together knowledge on an equal platform, using parallel systems of valuing and assessing knowledge, and Phase 3 is the joint analysis and evaluation of knowledge and insights to generate multilevel synthesis and identify and catalyze processes for generating new knowledge

systems in the current situation of pervasive environmental change and globalized societies (Krupnik and Jolly 2002).

Other scholars have claimed that valuable lessons can be extracted from such systems for complex system management, lessons that also need to account for interactions across temporal and spatial scales and organizational and institutional levels (Pretty and Ward 2001), in particular during periods of rapid change, uncertainty, and system reorganization (Berkes and Folke 2002).

2.4.2 *The Adaptive Cycle*

The adaptive cycle, originally conceptualized by Holling (1986), interprets the dynamics of complex ecosystems in response to perturbations and change. In terms of its dynamics, the adaptive cycle has been described as moving slowly from exploitation (r) to conservation (K), maintaining and developing very rapidly from K to release (W), and continuing rapidly to reorganization (a) and back to exploitation (r) (Gunderson and Holling 2002).

Depending on the particular configuration of the system, it can then begin a new adaptive cycle or alternatively it may transform into a new configuration, shown as an exit arrow (see Fig. 2.6). The adaptive cycle is one of the five heuristics used to understand social–ecological system behavior (Walker et al. 2006), the other four heuristics being resilience, panarchy, transformability, and adaptability. Each of these concepts is of considerable conceptual appeal and is claimed to be generally applicable to ecological and social systems as well as to coupled social–ecological systems (Gunderson and Holling 2002).

The two main dimensions that determine changes in an adaptive cycle are connectedness and potential (Gunderson and Holling 2002). The connectedness dimension is the visual depiction of a cycle and stands for the ability to internally control its own destiny (Holling 2001). It "reflects the strength of internal connections that mediate and regulate the influences between inside processes and the outside world"

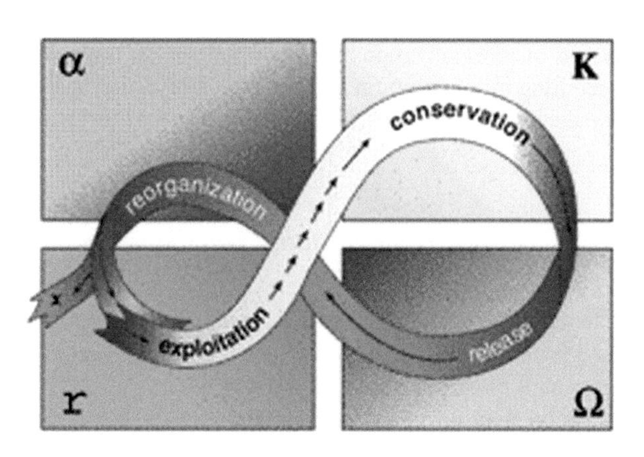

Fig. 2.6 Panarchy—graphical representation

(Gunderson and Holling 2002, p. 50). The potential dimension is represented by the vertical axis, and stands for the "inherent potential of a system that is available for change" (Holling 2001, p. 393). Social or cultural potential can be characterized by the "accumulated networks of relationships-friendship, mutual respect, and trust among people and between people and institutions of governance" (Gunderson and Holling 2002, p. 49). According to the adaptive cycle heuristic, the levels of both dimensions differ during the course of the cycle along the four phases. The adaptive cycle thus predicts that the four phases of the cycle can be distinguished based on distinct combinations of high or low potential and connectedness.

2.4.3 Adaptive Governance and SES

The resilience of social–ecological systems is related to the degree of the shock that the system can absorb and remain within a given state (Evans 2011). The concept of resilience is a promising tool for analyzing adaptive change towards sustainability because it provides a way for analyzing how to manipulate stability in the face of change.

In order to emphasize the key requirements of a social–ecological system for successful adaptive governance, Folke and colleagues (Folke et al. 2002) contrasted case studies from the Florida Everglades and the Grand Canyon. Both are complex social–ecological systems that have experienced unwanted degradation of their ecosystem services but differ substantially in terms of their institutional make-up.

The governance structure in the Everglades is dominated by the interests of agriculture and environmentalists who have been in conflict over the need to conserve the habitat at the expense of agricultural productivity throughout history. Here, a few feedbacks between the ecological system and the social system exist, and SES is unable to innovate and adapt (the α-phase of reorganization and growth).

In contrast, different stakeholders have formed an adaptive management workgroup in the case of Grand Canyon, using planned management interventions and monitoring to learn about changes occurring in the ecosystem, including the best ways to subsequently manage them. Such an arrangement in governance creates the opportunity for institutional learning to take place, allowing for a successful period of reorganization and growth. Such an approach to institutional learning is becoming more common as NGOs, scientist, and communities collaborate to manage ecosystems (Evans 2011).

2.4.4 Links to Sustainable Development

The concept of social–ecological systems has been developed in order to provide both promising scientific gains as well as to impact problems of sustainable development. A close conceptual and methodological relation exists between the analysis of social–ecological systems, complexity research, and transdisciplinarity. These three research concepts are based on similar ideas and models of reasoning. Moreover, the research on social–ecological systems almost always uses a transdisciplinary approach in order to achieve and ensure integrative results. Problems of sustainable development are intrinsically tied to the social–ecological system defined to tackle them. This means that scientists from the relevant scientific disciplines or field of research as well as the involved societal stakeholders have to be regarded as elements of the social–ecological system in question.

2.5 Earth System Dynamics and the Syndromes Approach

2.5.1 Syndromes Concepts

The main idea behind the "syndromes" approach (see Petschel-Held et al. 1995) is to couple the dynamics of ecosphere and anthroposphere.

The "holistic" approach proposed by Schellnhuber, Petschel-Held, and their colleagues (Schellnhuber et al. 1997; Petschel-Held et al. 1999; Petschel-Held and Reusswig 1999) aims at considering the Earth System itself as a sort of "system of systems," where the massive use of simulations of social- and ecological dynamics may help us to better understand the complex behavior of our planet.

The main criticism of different approaches (mainly, those considering models of the Earth System where the main effort is trying to reproduce and mimic micro-behaviors and properties of different components, more than dynamics) is that "analogous modeling by reproduction of the quantitative actual structure of the system may gain forecasting and hindcasting power only when the degree of sophistication becomes excessive" (Schellnhuber et al. 1997) (i.e., which, an exact copy of the real system of course, is scientifically quite useless).

The "syndromes" approach postulates that the overall phenomenon "Global Change" should be investigated as a co-evolutionary process of dynamic partial patterns and that these patterns "are bundles of interactive processes that are widespread and appear repeatedly in typical combination—the syndromes of global change".

In this perspective, syndromes are not simple complexes of causes and effects; they are patterns of interactions, frequently presenting feedbacks (see the concept of emerging cooperative phenomena in complex systems science).

Syndromes have a clear qualitative identity that cannot be quantified or measured using algorithms, metrics, or values. Because of this "soft identity" of syndromes and their interdisciplinary composition (syndromes are "active zones" of problematic environmental and development processes, rather than static patterns), we need specific and sometimes innovative methods of investigation, such as decomposition of complex functional networks, qualitative reasoning concepts, modeling of fuzziness and uncertainty, knowledge-acquisition strategies, and set-values analysis.

We have to take into account about 80 operating symptoms in doing a diagnosis of Earth System syndromes (contributing to the presence of different syndromes), such as urban sprawl, increasing significance of NGOs, terrestrial run-off changes deposition and accumulation of waste, increasing mobility, tropospheric pollution, and increasing consumption of energy and resources.

The names of these symptoms have to be interpreted as guiding headlines and not as definitions; they concern different spheres (e.g., atmosphere, biosphere, anthroposphere) and focus on qualitative and quantitative changes of the Earth System.

For Global Change analysis purposes the simple identification of symptoms is not sufficient; what is also crucial is the way they interact with each other. Such interactions have one "target symptom" and one or more "source symptoms" representing the causal connections between the symptoms involved.

The symptoms metaphor represents a dynamic and trans-disciplinary language to describe the Global Change phenomena. Symptoms indicate possible critical shifts towards nonsustainability. Since Global Change mainly refers to "anthropogenic" processes, symptoms usually are either direct expressions of human actions (for example, change of consumption patterns) or they are indirectly induced by it (for example, anthropogenic climate change). Thus, the micro–macro links, connections, and dynamics are critical for the syndromes approach; behaviors, habits, and interactions at the individual level may produce significant and critical effects at the global level (e.g., at the group, ecosystem, Earth System level).

The Earth System is not only a functional unit, it is a geographical one as well. This means that the correct use of the syndromes approach has to consider also the spatial scale of symptoms; otherwise there is a risk that important elements of the examined phenomenon may be missed.

2.6 Conclusions

This chapter presented an overview of some interesting applications of the agent-based modeling approach for investigating social behaviors and dynamics in living organisms; we also presented additional scientific domains (namely, the Social–Ecological Systems and Earth System Dynamics scientific fields) where ABM already represents or potentially could represent a very useful and promising approach.

Some of the main social behaviors and dynamics investigated in living organisms were presented—such as cooperation, information transmission and communicative systems, culture and imitation—in an effort to highlight the crucial role played by cognitive mechanisms and processes (e.g., social cognition).

ABM was analyzed as a scientific method and tool, and the argument was made that it may represent the best approach to dealing with complex nonlinear phenomena and dynamics.

References

Adger, N. (2000). Social and ecological resilience: Are they related? *Progress in Human Geography, 24,* 347–364.

Alexander, R. D. (1987). *The biology of moral systems*. New York: Aldine de Gruyter.

Alexander, J. C., & Giesen, B. (1987). From reduction to linkage: The long view of the micro-macro link. In J. C. Alexander, B. Giesen, R. Munch, & N. J. Smelser (Eds.), *The micro-macro link* (pp. 1–42). Berkeley: University of California Press.

Allen, T. F. H., & Starr, T. B. (1982). *Hierarchy: Perspectives for ecological complexity*. Chicago: University of Chicago Press.

André, J. B., & Baumard, N. (2011). Social opportunities and the evolution of fairness. *Journal of Theoretical Biology, 289,* 128–135.

Axelrod, R. (1984). *The evolution of cooperation.* New York: Basic Books.

Axelrod, R., & Hamilton, W. D. (1981). The evolution of cooperation. *Science, 211,* 1390–1396.

Barclay, P. (2011). Competitive helping increases with the size of biological markets and invades defection. *Journal of Theoretical Biology, 281,* 47–55.

Bateson, G. (1979) Mind and nature: A necessary unit. http://www.oikos.org/mind&nature.htm.

Berkes, F. (1989) *Common property resources: Ecology and community-based sustainable development*. London: Belhaven Press.

Berkes, F. (1999). *Sacred ecology: Traditional ecological knowledge and management systems.* Philadelphia: Taylor & Francis.

Berkes, F., & Folke, C. (2002). *Back to the future: Ecosystem dynamics and local knowledge.* In L. H. Gunderson & C. S. Holling (Eds.), *Panarchy: Understanding transformations in human and natural systems* (pp. 121–146). Washington, D.C.: Island Press.

Berkes, F., Colding, J., & Folke, C. (2000). Rediscovery of traditional ecological knowledge as adaptive management. *Ecological Applications, 10,* 1251–1262.

Berkes, F., Colding, J., & Folke, C. (2001). *Linking social-ecological systems*. Cambridge: Cambridge University Press.

Berkes, F., Colding, J., & Folke, C. (2003). *Navigating social-ecological systems: Building resilience for complexity and change*. Cambridge: Cambridge University Press.

Boekhorst, I. J. A., & Hogeweg, P. (1994a). Effects of tree size on travelband formation in orang-utans: Data-analysis suggested by a model. In R. A. Brooks & P. Maes (Eds.), *Artificial life* (pp. 119–129). Cambridge: The MIT Press.

Boekhorst, I. J. A. te, & Hogeweg, P. (1994b). Self- structuring in artificial 'CHIMPS' offers new hypotheses for male grouping in chimpanzees. *Behaviour, 130,* 229–52.

Bonabeau, E. (2002). Agent-based modeling: Methods and techniques for simulating human systems. *Proceedings of the National Academy of Sciences of the USA, 99*(3), 7280–7287.

Braeuer, J., Call, J., & Tomasello, M. (2007). Chimpanzees really know what others can see in a competitive situation. *Animal Cognition, 10*(4), 439–448.

Bryson, J. J., Ando, Y., & Lehmann, H. (2007). Agent-based modelling as scientific method: A case study analysing primate social behaviour. *Philosophical Transactions of the Royal Society, B—Biological Sciences, 362*(1485), 1685–1698.

Bshary, R., & Bronstein, J. L. (2011). A general scheme to predict partner control mechanisms in pairwise cooperative interactions between unrelated individuals. *Ethology, 117,* 271–283.

Bugnyar, T., & Heinrich, B. (2005). Ravens, Corvus corax, differentiate between knowledgeable and ignorant competitors. *Proceedings of the Royal Society, B—Biological Sciences, 272,* 1641–1646.

Bugnyar, T., & Heinrich, B. (2006). Pilfering ravens, Corvus corax, adjust their behaviour to social context and identity of competitors. *Animal Cognition, 9*(4), 369–376.

Bull, J. J., & Rice, W. R. (1991). Distinguishing mechanisms for the evolution of cooperation. *Journal of Theoretical Biology, 149,* 63–74.

Call, J., & Tomasello, M. (2008). Does the chimpanzee have a theory of mind? 30 years later. *Trends in Cognitive Sciences, 12*(5), 187–192.

Campennì, M., & Schino, G. (2014). Partner choice promotes cooperation: The two faces of testing with agent-based models. *Journal of Theoretical Biology, 344,* 49–55.
Cangelosi, A. (2001). Evolution of communication and language using signals, symbols, and words. *IEEE Transactions in Evolution Computation, 5*(2), 93–101.
Carpenter, S. R., & Gunderson, L. H. (2001). Coping with collapse: Ecological and social dynamics in ecosystem management. *BioScience, 51,* 451–457.
Cavalli-Sforza, L. L., Minch, E., & Mountain, J. L. (1992). Coevolution of genes and languages revisited. *Proceedings of the National Academy of Sciences of the USA, 89*(12), 5620–5624.
Charrier, I., & Sturdy, C. B. (2005). Call-based species recognition in the black-capped chickadees. *Behavioural Processes, 70*(3), 271–281.
Cheney, D. L. (2011). Extent and limits of cooperation in animals. *Proceedings of the National Academy of Science USA, 108,* 10902–10909.
Cheney, D. L., & Seyfarth, R. M. (1990). *How monkeys see the world: Inside the mind of another species*. Chicago: University of Chicago Press.
Clayton, N. S., Dally, J. M., & Emery, N. J. (2007). Social cognition by food-caching corvids. The western scrub-jay as a natural psychologist. *Philosophical Transactions of the Royal Society, B—Biological Sciences, 362*(1480), 507–522.
Clutton-Brock, T. H. (2009). Cooperation between non-kin in animal societies. *Nature, 462,* 51–57.
Costanza, R., Low, B. S., Ostrom, E., & Wilson, J. (2001). *Institutions, ecosystems, and sustainability*. Boca Raton: Lewis.
Cumming, G. S. (2011). *Spatial resilience in social-ecological systems*. London: Springer.
Dally, J. M., Emery, N. J., & Clayton, N. S. (2005). Cache protection strategies by western scrub-jays, Aphelocoma californica: Implications for social cognition. *Animal Behaviour, 70*(6), 1251–1263.
Dally, J. M., Emery, N. J., & Clayton, N. S. (2006). Food-caching western scrub-jays keep track of who was watching when. *Science, 312*(5780), 1662–1665.
Darwin, C. (1964). *The origin of species*. Cambridge: Harvard University Press (reprinted).
DeAngelis, D. L., Godbout, L., & Shuter, B. J. (1991). An individual-based approach to predicting density-dependent dynamics in smallmouth bass populations. *Ecological Modelling, 57*(1–2), 91–115.
Doebeli, M., & Hauert, C. (2005). Models of cooperation based on the prisoner's dilemma and the snowdrift game. *Ecology Letters, 8*(7), 748–766.
Dugatkin, L. A. (1997). *Cooperation among animals: An evolutionary perspective*. Oxford: Oxford University Press.
Dunbar, R. I. M., & Shultz, S. (2007). Evolution in the social brain. *Science, 317,* 1344–1347.
Dyer, F. C., & Seeley, T. D. (1991). Dance dialects and foraging range in three asian honey bee species. *Behavioral Ecology & Sociobiology, 28*(4), 227–233.
Emery, N. J., & Clayton, N. S. (2001). Effects of experience and social context on prospective caching strategies by scrub jays. *Nature, 414,* 443–446.
Epstein, J. M., & Axtell, R. (1996). *Growing artificial societies. Social science from the bottom up*. Washington, D.C.: Brookings Institution Press and Cambridge, MIT Press.
Eshel, I., & Cavalli-Sforza, L. L. (1982). Assortment of encounters and evolution of cooperativeness. *Proceedings of the National Academy of Science USA, 79,* 1331–1335.
Evans, J. (2011). *Environmental governance*. London: Routledge.
Evans, C. S., Evans, C. L., & Marler, P. (1993). On the meaning of alarm calls: Functional reference in an avian vocal system. *Animal Behaviour, 46*(1), 23–38.
Fitch, W. T. (2008). Glossogeny and phylogeny: Cultural evolution meets genetic evolution. *Trends in Genetics, 24*(8), 373–374.
Fitch, W. T. (2010). *The evolution of language*. New York: Cambridge University Press.
Fitch, W. T., Huber, L., & Bugnyar, T. (2010). Social cognition and the evolution of language: Constructing a cognitive phylogeny. *Neuron, 65*(6), 795–814.
Folke, C. (2006). Resilience: The emergence of a perspective for social-ecological systems analysis. *Global Environmental Change, 16,* 253–267.

Folke, C., Carpenter, S., Elmqvist, T., Gunderson, L., Holling, C., & Walker, B. (2002). Resilience and sustainable development: Building adaptive capacity in a world of transformations. *Ambio, 31,* 437–440.

Frank, R. H. (1989). *Passions within reason*. New York: W.W. Norton & Company.

Fruteau, C., Lemoine, S., Hellard, E., Van Damme, E., & Noë, R. (2011). When females trade grooming for grooming: Testing partner control and partner choice models of cooperation in two primate species. *Animal Behaviour, 81,* 1223–1230.

Galef, B. G. J., & Laland, K. N. (2005). Social learning in animals: Empirical studies and theoretical models. *BioScience, 55*(6), 488–489.

Gilbert, N., & Troitzsch, K. (2005). *Simulation for the social scientist* (2nd edn). England: Open University Press.

Green, E., & Maegner, T. (1998). Red squirrels, Tamiasciurus hudsonicus, produce predator-class specific alarm calls. *Animal Behaviour, 55*(3), 511–518.

Greenberg, J. B., & Park, T. K. (1994). Political ecology. *Journal of Political Ecology, 1,* 1–12.

Gunderson, L. H., & Holling, C. S. (2002). *Panarchy: Understanding transformations in human and natural systems*. Washington, D.C.: Island Press.

Hailman, J., Ficken, M., & Ficken, R. (1985). The chick-a-dee calls of Parus atricapillus. *Semiotica, 56*(3–4):191–224.

Hamilton, W. D. (1963). The evolution of altruistic behaviour. *American Naturalist, 97*(896), 354–356.

Hamilton, W. D. (1964). The genetical evolution of social behaviour. *Journal of Theoretical Biology, 7*(1), 17–52.

Hammerstein, P. (Ed.). (2003). *Genetic and cultural evolution of cooperation.* Cambridge: MIT Press.

Hare, B. (2001). Can competitive paradigms increase the validity of experiments on primate social cognition? *Animal Cognition, 4*(3–4), 269–280.

Hare, B., Call, J., Agnetta, B., & Tomasello, M. (2000). Chimpanzees know what conspecifics do and do not see. *Animal Behaviour, 59*(4), 771–785.

Hauert, C. (2001). Fundamental clusters in spatial 2×2 games. *Proceedings of the Royal Society, B—Biological Sciences, 268*(1468), 761–769.

Hemelrijk, C. K. (1996). Dominance interactions, spatial dynamics and emergent reciprocity in a virtual world. In P. Maes, M. J. Mataric, J.-A. Meyer, J. Pollack, & S. W. Wilson (Eds.), *Proceedings of the fourth international conference on simulation of adaptive behavior 4* (pp. 545–552). Cambridge: MIT-Press.

Hemelrijk, C. K. (2000). Towards the integration of social dominance and spatial structure. *Animal Behaviour, 59*(5), 1035–1048.

Heyes, C. M. (1994). Social learning in animals: Categories and mechanisms. *Biological Reviews, 69*(2), 207–231.

Heyes, C. (2009). Evolution, development and intentional control of imitation. *Philosophical Transactions of the Royal Society, B—Biological Sciences, 364*(1528), 2293–2298.

Hoffman, M., Yoeli, E., & Nowak, M. A. (2015). Cooperate without looking: Why we care what people think and not just what they do. *Proceedings of the National Academy of Sciences, 112*(6), 1727–1732.

Holling, C. S. (1986). The resilience of terrestrial ecosystems: Local surprise and global change. *Sustainable Development of The Biosphere,* 292–317.

Holling, C. S. (2001). Understanding the complexity of economic, ecological, and social systems. *Ecosystems, 4*(5), 390–405.

Huber, L., Range, F., Voelkl, B., Szucsich, A., Viranyi, Z., & Miklosi, A. (2009). The evolution of imitation: What do the capacities of nonhuman animals tell us about the mechanisms of imitation? *Philosophical Transactions of the Royal Society, B—Biological Sciences, 364*(1528), 2299–2309.

Johnstone, R. A., & Bshary, R. (2002). From parasitism to mutualism: Partner control in asymmetric interactions. *Ecology letters, 5,* 634–639.

Kaiser, D. (2004). Signaling in myxobacteria. *Annual Review of Microbiology, 58,* 75–98.

Kaminski, J., Call, J., & Tomasello, M. (2008). Chimpanzees know what others know, but not what they believe. *Cognition, 109*(2), 224–234.
Karin-D'Arcy, M., & Povinelli, D. J. (2002). Do chimpanzees know what each other see? A closer look. *International Journal of Comparative Psychology, 15,* 21–54.
Killingback, T., Doebeli, M., & Knowlton, N. (1999). Variable investment, the continuous prisoner's dilemma, and the origin of cooperation. *Proceedings of the Royal Society, B—Biological Sciences, 266*(1430), 1723–1728.
Kim, J. (1992). "Downward causation" in emergentism and non-reductive physicalism. In A. Beckermann, H. Flohr, & J. Kim (Eds.), *Emergence or reduction? Essays on the prospects of nonreductive physicalism*. Berlin: Walter de Gruyter.
Kreft, J. U., Booth, G., & Wimpenny, J. W. T. (1998). Bacsim, a simulator for individual-based modelling of bacterial colony growth. *Microbiology, 144*(12), 3275–3287.
Krupnik, I., & Jolly, D. (2002). *The earth is faster now: Indigenous observation on arctic environmental change*. Fairbanks: Arcus.
Leadbeater, E., & Chittka, L. (2007). Social learning in insects—from miniature brains to consensus building. *Current Biology, 17*(16), 703–713.
Lehmann, L., & Keller, L. (2006). The evolution of cooperation and altruism. A general framework and classification of models. *Journal of Evolutionary Biology, 19,* 1365–1376.
Levin, S. A. (1998). Ecosystems and the biosphere as complex adaptive systems. *Ecosystems, 1,* 431–436.
Levin, S. A. (1999). *Fragile dominion: Complexity and the commons*. Reading: Perseus Books.
Lieberman, E., Michel, J.-B., Jackson, J., Tang, T., & Nowak, M. A. (2007). Quantifying the evolutionary dynamics of language. *Nature, 449,* 713–716.
Machlis, G. E., Force, J. E., & Burch, W. R. Jr. (1997). The human ecosystem part I: The human ecosystem as an organizing concept in ecosystem management. *Society and Natural Resources, 10,* 347–367.
Mackinson, S., & Nottestad, L. (1998). Combining local and scientific knowledge. *Reviews in Fish Biology and Fisheries, 8,* 481–490.
Manser, M., Seyfarth, R. M., & Cheney, D. L. (2002). Suricate alarm calls signal predator class and urgency. *Trends in Cognitive Sciences, 6*(2), 55–57.
McCay, B. J., & Acheson, J. M. (1987). *The question of the cotntnons. The culture and ecology of comtnunal resources*. Tucson: The University of Arizona Press.
McLain, R., & Lee, R. (1996). Adaptive management: Promises and pitfalls. *Journal of Environmental Management, 20,* 437–448.
Mesoudi, A., Whiten, A., & Laland, K. N. (2004). Perspective: Is human cultural evolution darwinian? Evidence reviewed from the perspective of the origin of species. *Evolution, 58*(1), 1–11.
Gilbert, N. (2008). *Agent-based models (quantitative applications in the social sciences)*. Thousand Oaks: Sage Publications, Inc.
Noble, J. (1999). Cooperation, conflict and the evolution of communication. *Adaptive Behavior, 7*(3–4), 349–369.
Noble, J., de Ruiter, J. P., & Arnold, K. (2010). From monkey alarm calls to human language: How simulations can fill the gap. *Adaptive Behavior, 18*(1), 66–82.
Noë, R. (2006). Cooperation experiments: Coordination through communication versus acting apart together. *Animal Behaviour, 71,* 1–18.
Noë, R., & Hammerstein, P. (1994). Biological markets: Supply and demand determine the effect of partner choice in cooperation, mutualism and mating. *Behavioral Ecology and Sociobiology, 35,* 1–11.
Norberg, J., & Cumming, G. S. (2008). *Complexity theory for a sustainable future*. New York: Columbia University Press.
Nowak, M. A. (2006). Five rules for the evolution of cooperation. *Science, 314,* 1560–1563.
Nowak, M. A., & May, R. M. (1992). Evolutionary games and spatial chaos. *Nature, 359,* 826–829.
Nowak, M. A., & Sigmund, K. (1998). Evolution of indirect reciprocity by image scoring. *Nature, 393,* 573–577.
Nowak, M. A., Tarnita, C. E., & Wilson, E. O. (2010). The evolution of eusociality. *Nature, 466,* 1057–1062.

Nunn, C. L., & Lewis, R. J. (2001). Cooperation and collective action in animal behavior. In R. Noë, van J. A. R. A. M. Hooff, & P. Hammerstein (Eds.), *Economics in nature* (pp. 42–66). Cambridge: Cambridge University Press.
Pagel, M., Atkinson, Q. D., & Meade, A. (2007). Frequency of word-use predicts rates of lexical evolution throughout indo-european history. *Nature, 449,* 717–720.
Paolucci, M., Conte, R., & Di Tosto, G. (2006). A model of social organization and the evolution of food sharing in vampire bats. *Adaptive Behavior, 14*(3), 223–239.
Petschel-Held, G., et al. (1999). Syndromes of global change—A qualitative modelling approach to assist global environmental management. *Environmental Modeling and Assessment, 4,* 295–314.
Petschel-Held, G., & Reusswig, F. (1999). Climate change and global change—The syndrome concept. In J. Hacker & A. Pelchen (Eds.), *Goals and economic instruments for the achievement of global warming mitigation in Europe* (pp. 79–95). Dordrecht: Kluwer.
Petschel-Held, G., Block, A., & Schellnhuber, H.-J. (1995). Syndrome des Globalen Wandels—ein systemarer Ansatz für Sustainable-Development-Indikatoren. *GEOwissenschaften, 3,* 81–87.
Povinelli, D. J., & Eddy, T. J. (1996). Chimpanzees: Joint visual attention. *Psychological Science, 7*(3), 129–135.
Povinelli, D. J., & Vonk, J. (2003). Chimpanzee minds: Suspiciously human? *Trends in Cognitive Sciences, 7*(4), 157–160.
Povinelli, D. J., Nelson, K. E., & Boysen, S. T. (1990). Inferences about guessing and knowing by chimpanzees (Pan troglodytes). *Journal of Comparative Psychology, 104*(3), 203–210.
Premack, D., & Woodruff, G. (1978). Does the chimpanzee have a theory of mind? *Behavioral and Brain Sciences, 1*(4), 515–526.
Pretty, J., & Ward, H. (2001). Social capital and the environment. *World Development, 29,* 209–227.
Redman, C., Grove, M. J., & Kuby, L. (2004). Integrating social science into the Long Term Ecological Research (LTER) Network: Social dimensions of ecological change and ecological dimensions of social change. *Ecosystems, 7*(2), 161–171.
Reynolds, C. W. (1987). Flocks, herds, and schools: A distributed behavioral model. *Computer Graphics, 21*(4), 25–34 (SIGGRAPH '87 Conference Proceedings).
Roberts, G. (1998). Competitive altruism: From reciprocity to the handicap principle. *Proceedings of the Royal Society B, 265,* 427–431.
Sachs, J. L., Mueller, U. G., Wilcox, T. P., & Bull, J. J. (2004). The evolution of cooperation. *The Quarterly Review of Biology, 79,* 135–160.
Schauder, S., & Bassler, B. L. (2001). The languages of bacteria. *Genes & Development, 15,* 1468–1480.
Schellnhuber, H.-J., et al. (1997). Syndromes of global change. *GAIA, 6,* 19–34.
Schermerhorn, P., & Scheutz, M. (2007). Investigating the adaptiveness of communication in multi-agent behavior coordination. *Adaptive Behavior, 15*(4), 423–445.
Schino, G. (2007). Grooming and agonistic support: A meta-analysis of primate reciprocal altruism. *Behavioural Ecology, 18,* 115–120.
Schino, G., & Aureli, F. (2008). Grooming reciprocation among female primates: A meta-analysis. *Biology Letters, 4,* 9–11.
Schino, G., & Aureli, F. (2009). Reciprocal altruism in primates: Partner choice, cognition, and emotions. *Advances in the Study of Behavior, 39,* 45–69.
Schuster, S., Woehl, S., Griebsch, M., & Klostermeier, I. (2006). Animal cognition: How archer fish learn to down rapidly moving targets. *Current Biology, 16*(4), 378–383.
Seyfarth, R. M., & Cheney, D. L. (1990). The assessment by vervet monkeys of their own and other species' alarm calls. *Animal Behaviour, 40*(4), 754–764.
Seyfarth, R. M., & Cheney, D. L. (2012). The evolutionary origins of friendship. *The Annual Review of Psychology, 63,* 153–177.
Sherratt, T. N., & Roberts, G. (1998). The evolution of generosity and choosiness in cooperative exchanges. *Journal of theoretical biology, 193,* 167–177.
Sigmund, K., Hauert, C., & Nowak, M. A. (2001). Reward and punishment. *Proceedings of the National Academy of Sciences of the USA, 98*(19), 10757–10762.

Sigmund, K., Fehr, E., & Nowak, M. A. (2002). The economics of fair play. *Scientific American, 286,* 82–87.
Skyrms, B. (2009). Evolution of signalling systems with multiple senders and receivers. *Philosophical Transactions of the Royal Society B—Biological Sciences, 364*(1518), 771–779.
Smith, K. (2002). Natural selection and cultural selection in the evolution of communication. *Adaptive Behavior, 10*(1), 25–45.
Sober, E., & Wilson, D. S. (1998). *Unto others: The evolution and psychology of unselfish behavior*. Cambridge: Harvard University Press.
Striedter, G. F. (2004). *Principles of brain evolution*. Sunderland: Sinauer.
Taga, M. E., & Bassler, B. L. (2003). Chemical communication among bacteria. *Proceedings of the National Academy of Sciences of the USA, 100*(2), 14549–14554.
Tennie, C., Call, J., & Tomasello, M. (2009). Ratcheting up the ratchet: On the evolution of cumulative culture. *Philosophical Transactions of the Royal Society B—Biological Sciences, 364*(1528), 2405–2415.
Tiddi, B., Aureli, F., Polizzi di Sorrentino, E., Janson, C. H., & Schino, G. (2011). Grooming for tolerance? Two mechanisms of exchange in wild tufted capuchin monkeys. *Behavioural Ecology, 22,* 663–669.
Tomasello, M., Call, J., & Hare, B. (2003). Chimpanzees understand psychological states—The question is which ones and to what extent. *Trends in Cognitive Sciences, 7*(4), 153–156.
Trivers, R. L. (1971). The evolution of reciprocal altruism. *The Quarterly Review of Biology, 46,* 35–57.
Wagner, K., Reggia, J. A., Uriagereka, J., & Wilkinson, G. S. (2003). Progress in the simulation of emergent communication and language. *Adaptive Behavior, 11*(1), 37–69.
Walker, B. H., Gunderson, L. H., Kinzig, A. P., Folke, C., Carpenter, S. R., & Schultz, L. (2006). A handful of heuristics and some propositions for understanding resilience in social-ecological systems. *Ecology and Society, 11*(1), 13. http://www.ecologyandsociety.org/vol11/iss1/art13/.
Warren, D. M., Slikkerveer, L. J., & Brokensha, D. (1995). *The cultural dimension of development: Indigenous knowledge system*. London: Intermediate Technology Publications.
West, S. A., Griffin, A. S., & Gardner, A. (2007). Evolutionary explanations for cooperation. *Current Biology, 17,* 661–672.
Wilkinson, A., Kuenstner, K., Mueller, J., & Huber, L. (2010). Social learning in a non-social reptile (Geochelone carbonaria). *Biology Letters, 6*(5), 614–616.
Wilson, E. O. (1975). *Sociobiology*. Cambridge: Harvard University Press.
Wilson, D. S., & Sober, E. (1994). Reintroducing group selection to the human behavioral sciences. *Behavioral & Brain Sciences, 17*(4), 585–654.
Zentall, T. R. (2004). Action imitation in birds. *Learning & Behavior, 32*(1), 15–23.

Chapter 3
Reciprocity, Punishment, Institutions: The Streets to Social Collaboration—New Theories on How Emerging Social Artifacts Control Our Lives in Society

Giuliana Gerace

3.1 Introduction

Cooperation studies are extremely important for real-life purposes, e.g., economic predictions and policy design, but they are especially fascinating for their broad theoretical implications.

In the last few decades a flourishing body of investigations concerning the emergence of social collaboration has demonstrated the power of "reciprocity" in inducing cooperation, while demonstrating a heavy influence of pro-social behavior in social interactions, as opposed to the traditional conception of self-interested *homo oeconomicus*. A wealth of experimental evidence has demonstrated that cooperation occurs even when it is not predicted by economic theory, in contrast to the neoclassical assumption that the narrow pursuit of interest results in efficient economic exchanges.

The behavioral relevance of some social contextual determinants of reciprocity, such as reputation, altruistic punishment, and trust, showed that even short-run altruism, independent of contingent material payoff, can be regarded as rational in the long run and eventually be established as a behavioral norm in society. It also demonstrated that all kinds of social relations, including business relations, may rely on social binding conditions that are shared among individuals.

Evolutionary approaches in social science and game theory have investigated how cooperation can be induced by supporting mechanisms which limit the costs of incomplete information in bounded rationality conditions and basically uncertain environments of interaction. It is now thought that building behavioral decisions not on random events but on long-term accumulation of "social capital" is an evolutionarily rational behavior that reduces the possibility of uncertain interactions.

An optimal design of institutions or contracts, however, is highly difficult to conceive, especially in large groups, where monitoring and sanctioning solutions

G. Gerace (✉)
Department of Philosophy, University of Pavia, Piazza Botta 6, Pavia, Italy
e-mail: giuliana.gerace@gmail.com

F. Cecconi (ed.), *New Frontiers in the Study of Social Phenomena*,
DOI 10.1007/978-3-319-23938-5_3

require implausible cognitive capacities and high costs of coordination. In this case the cost–benefit valence of coordination is theoretically nullified by the costs for efficiently aligning individuals' behavior.

How then to justify so much cooperation in social interaction? What mechanisms induce and support such a strong convergence toward common, often articulated behavioral standards, and what are the factors which allow such standards to become resilient and/or stable in social interactions?

Notwithstanding its robustness, the weak explanatory potential of standard rational- choice theory (Sugden 1991) can hardly be invoked in this regard. Given the rigid assumptions that individuals are manifestly utility maximizers and always rely on complete information for decisions and lacking also any specification of the notion of a "utility function" as the basis of individual preferences, the standard theory faces problems in accounting for individual costly pro-sociality. In addition, it is rather ineffective to keep track of complex, real-world social interactions which spontaneously articulate in long-term shared obligations and offer everything but complete decision models, but rather are subject to disequilibrium and environmental changes.

Models and experimental studies on the emergence of cooperation therefore focus on adaptive or learning strategies that can be implemented through finite cognitive abilities; meanwhile they keep track of individual psychological attitudes in the social sphere, with a view to justifying non-standard rationality of social agents.

Nonetheless, whereas the perspective of traditional rational-choice theory is robust but inaccurate, perspectives based on the justification of fairness preferences as internalized behavioral forces driving realistic cooperative interactions are notoriously incomplete and rather fuzzy with respect to their theoretical foundations. They especially fail to give an adequate account of internalization processes and of the alleged interplay between cognitive and motivational factors responsible for individual social engagements.

On these bases, there is the urgent need of alternative more effective theoretical grounds for the emergence and variation of social conformity, able to account for individuals' convergence upon both shared norm compliance and shared behavioral dispositions.

In the following sections we will take into consideration widely recognized accounts from evolutionary approaches in social science and game theory, focusing on how standards of cooperation and coordination emerge. In addition we will consider extremely convincing accounts of how such standards can evolve as autonomous entities and eventually be stabilized in institutional forms.

At the same time, we will also focus on a perspective according to which the key to understanding evolutionary dynamics of social engagement is to be found in individual motivational attitudes to interaction, which may provide consistent justification without the need to rely on psychological implication. To be precise, we will suggest not exiting from the "logic of reciprocity" in considering individual rationality of preferences for social interaction as basically conditional to salience in social contexts. Finally, we will provide preliminary supporting experimental evidence.

3.2 The Emergence of Cooperation

3.2.1 Reciprocity Mechanisms

Reciprocity is a mutual condition of relationship, possibly based on cooperative or non-cooperative interchange. Sociologists have considered reciprocity as a sort of golden rule of interactions (Gouldner 1960) and this perspective has been strongly employed in the study of sociality among non-human animals (De Waal 1996).

Social scientists have studied reciprocity widely as an evolutionary factor promoting cooperation (Axelrod 1984). A preliminary explanation for the reasons that natural selection equipped selfish individuals with altruistic tendencies in reciprocal interactions is provided by the theory of kin selection, which considers much of human positive reciprocation to be driven by kinship (Hamilton 1964). Later on seminal works about the notion of reciprocal altruism (Trivers 1971) allowed researchers to study the emergence of cooperation among individuals without relying on kinship or fellowship, but just on their prior interactions (Axelrod and Hamilton 1981). Such forms of cooperation, emerging both in view of and by virtue of other individuals' direct collaboration, is generally regarded as direct reciprocity. Altruistic actions are performed as long as there is some expectation of future reciprocal cooperation. In this context, reciprocity can be seen as a strategic interaction between unrelated individuals and groups of individuals and also qualifies as a coordination device. As such, direct reciprocity proves important at all levels of social exchanges, influencing negotiations in conflicts, bargaining in international settings, and compliance in more restricted economic and political scenarios (Cialdini 1993).

Economists and game-theorists have widely used the notion of direct reciprocity to study the emergence of cooperation via social dilemmas. Particularly, in the common Prisoner Dilemma, players are basically led to defect, yet they still can choose to collaborate since they know that mutual cooperation results in a better outcome than mutual defection (as standard rational choice suggests)—hence the dilemma. In strategic interactions of this kind, reciprocal altruists always take somewhat of a risk: they have to rely on the goodwill (and good memory) of the recipient to return the favor.

A positive direct reciprocation must therefore assume (i) a random unknown number of repeated interactions (to avoid optimal calculus by defectors); (ii) a cost for the altruist that is inferior to the benefit provided to the recipient (b – c), since the rewards may not be reciprocated; (iii) sufficient cognitive attitudes for players to identify one another and accurately recall previous interactions (in order to gain trusting/non-trusting expectations).

Laboratory simulations of iterated Prisoner Dilemma strategies are only apparently winning solutions for cooperation. For example, in so-called tit-for-tat, individuals always end up collaborating by cooperating in the first tournament and then acting exactly as the opponent did in previous rounds; but indeed this expresses an unrealistic determinism (Axelrod and Dion 1988): real-life cooperation also emerges within initially uncooperative situations and especially in dynamic environments of interaction, affected by noise and changing choices.

Also on these bases, further investigations led to the notion of indirect reciprocity (Alexander 1987; Nowak and Sigmund 2005; Rockenbach and Milinski 2006; Wedekind and Milinski 2000), according to which individuals can receive long-term benefits for their short-term pro-social behavior, in this case not from recipients but from third parties. Players identify and possibly trust one another, not by recalling previous interactions, but rather by relying on acquaintanceship with opponents' general behaviors.

The potential of indirect reciprocity clearly involves the role of reputation: it is not required for two individuals to have ever met in order to cooperate; what matters is individuals' image as cooperators within the community (Nowak and Sigmund 1998; Wedekind and Milinski 2000). In this context language and "gossip" can be main vehicles of reputation (Alexander 1987; Nowak and Sigmund 1998; Panchanathan and Boyd 2003), while so-called reputation building can be regarded as an investment for future returns: I help you and probably other people will help me.

In this sense reputation is a sort of basic commodity (Ohtsuki et al. 2009), able to generate a widespread social income (i.e. cooperation). Costly acts may also help this endeavor (Zahavi and Zahavi 1997; Gintis et al. 2001), since observable acts of altruism, although costly, may establish individuals' positive images.

The interesting theoretical challenge of indirect reciprocity (very welcomed by evolutionary game theory) is that it allows us to study the emergence of cooperation in groups where partners meet only once. Clearly, the question of whether cooperation can ever be sustained under one-shot interactions is an interesting theoretical one. Particularly, whether one-shot interactions can stably sustain mutual cooperation based on minimal forms of reputation building has been the subject of considerable debate (Uchida and Sigmund 2010).

Image-scoring strategy (Nowak and Sigmund 1998) presents a key weakness, which renders it evolutionary unstable (Panchanathan and Boyd 2003). If help is preferentially directed toward recipients with a positive reputation, defectors are penalized, but discriminators who refuse to help recipients with a bad score receive bad scores and risk to be discriminated in turn. In this sense, punishing defectors by withholding help is a costly and non-evolving trait.

A more advantageous and non-costly strategy would be the standing rule (Sugden 1986; Ohtsuki and Iwasa 2006), distinguishing between justifiable defections (against bad recipients) and non-justifiable defections (against good recipients), and attaching bad scores only to the latter. Nonetheless, as well as being more evolutionarily stable than image scoring, since it relies on costless and truthful reputation building, standing strategy requires higher-order assessment rules, i.e., cognitively highly demanding solutions. It is often maintained that, by being contingent not only on past actions, but also on such action targets reputation, the standing strategy requires individuals with an implausibly large capacity for processing recursive information and of observability in large groups (Kandori 1992; Milinsky et al. 2001; Brandt et al. 2006).

Interestingly, however, recent studies (Berger and Grüne 2014) revalued image-scoring as a promoter of more-or-less stable cooperation, by assuming a multi-valued model instead of a traditional binary model, and by assuming that adaptive agents update their network of acquaintances through more than just one observed past action.

Until now, there has been little evidence of observability power promoting large-scale cooperation in real-world settings. It is generally maintained that observability works in small and rigid networks of rational players, i.e., networks allowing for certain familiarity and membership among individuals. As recent studies confirm (Rezaei et al. 2009; Rezai and Kirley 2012), when agents increase their cognitive capacity to classify their environment, social links play an increasingly important role in promoting and sustaining cooperation. Interestingly, the dynamic adjustment of social links results in the formation of communities of "like-minded" cooperative agents. A similar phenomenon is observable in random clustering (Hauert and Szabò 2005).

Along the same lines, as will be analyzed more thoroughly in the next subsection, evidence of "strong reciprocity" in local contexts (Gintis 2000; Bowles and Gintis 2004) showed that individuals belonging to (small) social groups perform highly costly cooperative behaviors that are not dependent on rational strategies for equilibrium.

On these bases, as well as the idea that reciprocity mechanisms are leading evolutionary theories of human cooperation, different types of evolutionary dispositions have been studied in order to understand particularly how cooperators are led to outperform non-cooperators in large populations and are therefore favored by selection. We will mention some examples.

Spatial selection posits that spatially structured populations and local interaction lead agents to cluster both in physical space and in social networks (Hauert and Szabò 2005; Nowak et al. 2010). Group selection, or multilevel selection (Wilson 1975), focuses instead on competitive interaction between groups, with intergroup competition as a powerful force in promoting within-group cooperation. Cultural selection, on the other side, relates to the possibility that cultural similarity (including minimal cues of shared identity or group paradigms) promotes the emergence of long-term cooperation, increasing cooperation among strangers exactly as genetic similarity does. With respect to effects, then, cultural-based cooperation performs as kin-based cooperation (Sigmund and Nowak 2001).

3.2.2 *Behavioral Patterns: Strong Reciprocity*

Besides mechanisms for cooperation, evolutionary approaches usually focus on the role of specific behavioral patterns not directly affecting the evolution of cooperation but able to increase and stabilize its level (Rand and Nowak 2013).

Important examples in this regard are upstream reciprocity, where helped individuals are more likely to be helpful in turn (Nowak and Roch 2006) and parochial reciprocity, according to which individuals are more likely to help individuals belonging to their own group than members of other groups (Bernhard et al. 2006).

The most widely studied pattern is strong reciprocity: individuals reward collaborators and incur costs to punish uncooperative individuals, without tangible individual benefits (Gintis 2000; Milinski et al. 2001; Bowles and Gintis 2004). A certain robustness and frequency of strong reciprocity has been observed across different cultures (Henrich et al. 2001; Gächter and Herrmann 2009). Evolutionary game theorists have accounted for evidence of pro-social behavior among non-kin individuals in one-shot anonymous interactions, which significantly contrasts with standard rationality. Particularly, the fact that unfair offers are frequently rejected in so-called ultimatum games constitutes an important piece of experimental evidence for strong reciprocity (Fehr et al. 2002).

Current investigations focus on possible justifications of strong reciprocity, as it is hard to rationalize as an adaptive trait of human cooperation. How do we explain pro-sociality in one-shot anonymous settings where no mechanisms of cooperation are explicitly present?

In contrast to self-interested conditional cooperation in strategic interactions, also definable as weak reciprocity (Gintis et al. 2005), strong reciprocity is usually considered to be a more complex impulse of the individual toward cooperation, also requiring an unexpressed attachment to altruism or so-called social preferences (Fehr and Schmidt 1999).

Strong reciprocity is not unconditional altruism. It rather expresses an interest for a higher-order form of reciprocation: strong reciprocators stop cooperating with cheaters and punish them, exactly because they are strongly interested in some individuals' convergence upon shared behavioral standards. Namely, they are interested in some norm of cooperation. How to justify the emergence of such higher-order interest?

Indeed, strong reciprocity is consistent with selection theories of human cooperation. Different types of evolutionary forces, including costly signaling, can be plausibly responsible for individual strong motivations to engage in costly behaviors (Fehr and Fischenbar 2003; Bowles and Gintis 2011; Rand and Nowak 2013). But the question of whether such altruistic behavior, with no repetition or reputation effects, has to be explained by selection theories or by higher-order reciprocity mechanisms is still an open one.

A large amount of research demonstrates the power of reciprocity in inducing cooperation. Even selection mechanisms can be regarded as an effect of the most advantageous reciprocation strategies emerging in repeated interactions (Rand and Nowak 2013), which affect individuals' dispositions thanks to the powerful role of learning, imitation, and internalization processes (Bowles and Gintis 2011; Rand et al. 2012).

The notion of internalization (Gintis 2004; Bowles and Gintis 2011) is an important one in the context of social heuristics. Internalization occurs when individual motivation for norm-compliance stops relying on exogenous factors (e.g., sanction/reward) and begins to rely directly on what the norm stands for. Intuitively, a crucial factor of social norm internalization is the preference that other individuals also comply with the same norm.

It is reasonable to assume that while the environment of evolutionary adaptation creates the heuristics for playing repeated games efficiently, some cooperation strategies yielding higher payoffs are internalized as "social norms," and this may be a plausible explanation for the emergence of strong reciprocation.

An interesting aspect of strong reciprocation is the potential to affect other individuals with cooperative behavior: how can selfish types and strong reciprocators affect one another in interactions? The presence/absence of punishment opportunities seems to be crucial here (Fehr and Fischenbar 2003). Indeed, at least with regard to stabilizing cooperation, sanctions can be regarded as a viable solution. But as it will be seen, it is interesting how reciprocity mechanisms are capable of mutual enforcement building on cooperation itself. In fact, it has been observed that a common feature of successful models for cooperation is the positive assortment of altruists across time (Bowles and Gintis 2011, p. 48), which confirms that the very determinant of the evolutionary dynamics of cooperation is positive reciprocation. In this context, while the possibility of enhancing collective actions by means of norm enforcing is an important aspect of strong reciprocity patterns, on the other hand, the fact that people are more likely to cooperate if they observe/believe others are also cooperating is equally important trait.

3.2.2.1 Punishment

The threat of punishment can lead to considerable increases in the level and longevity of cooperation in social interaction (Gintis and Fehr 2012). Punishment is a specific behavioral attitude which can promote cooperation in different ways. Direct reciprocity punishment is an in-kind response to harmful acts (retaliation or negative reciprocity), while indirect reciprocity punishment is a form of naturally emerging sanction, i.e., discrimination. Costly punishment or altruistic punishment (Fehr and Gächter 2002), typical of strong reciprocity mechanisms, has attracted considerable attention because individuals voluntarily incur costs with no future tangible benefits. Here, as mentioned earlier, the canonical model of self-interested material payoff maximization is violated in order to maximize higher-order preferences (social preferences).

The incentive structure within which inter-group cooperation is maintained through altruistic punishment by strong reciprocators is called "self-policing." Experimental results show that, under appropriate conditions, altruistic punishment can sustain the maintenance of high levels of cooperation unless the frequency of strong reciprocators is too low or the group is too large: punishing is costly, and if the desire to punish is not sufficiently widespread, self-policing will fail (Carpenter et al. 2009). Also an interaction between punishment and reputation building boosts cooperative efficiency (Rockenbach and Milinski 2006): the costs of punishment are markedly reduced in association with the appreciation of another's reputation.

The main problem with self-policing is that it rests on uncoordinated forms of punishment. Under such conditions the sum of costs to punishers often exceeds the benefits of increased cooperation (Ohtsuki et al. 2009; Boyd et al. 2010). As a result, cooperation sustained by uncoordinated voluntary punishment reduces the average payoffs of group members in comparison with groups in which punishment of free-riders is not an option. In light of this thorny problem, it has been maintained that self-policing punishment, also defined as peer punishment, can be positively contrasted with so-called pool punishment—namely, the possibility for individuals to engage in some form of social contract in order to delegate punishment to a third party (Ostrom 2005). In nearly all developed, regulated societies peer punishment is explicitly forbidden on favor of specifically designated institutions which establish modalities and ongoing costs of pool punishment in advance.

Clearly, if altruistic punishment worked as desired, i.e., lead to all-out cooperation, peer punishment could be more efficient than pool punishment, with no need for ongoing costs to be incurred by the sanctioning structure. Nonetheless, since informal control in peer punishment rests on the possibility of repeated, non-anonymous interactions, vital factors of such decentralized control, (e.g. signaling, retaliation, and reputation formation) are socially undesirable as well as unattainable mechanisms in large communities.

A centralized punishment structure, instead, offers the advantages of coordination and higher order stability, whereby pool punishers can mutually enforce their support to the punishment structure, stably trapping each other (Sigmund et al. 2010; Zahng et al. 2014).

Experimental models of cost-effective coordinated sanctioning (Boyd et al. 2010; Bowles and Gintis 2011) show that institutionalized punishing is particularly advantageous: sharing costs of sanctioning influences higher-order community cooperation, such as rewards to punishers and free riders discrimination. Such mechanisms often work in conjunction with social emotions such as public shame (Bowles and Gintis 2011).

Experimental results also show the possibility of spontaneous emergence of pool punishment. Social learning can lead to individual preferences for coordination in matters of punishment, especially when sanctions are also imposed on second-order free-riders, namely, individuals in charge of punishment who don't accomplish their tasks (Zahng et al. 2014). The sanctioning system is regarded as a public good itself to be exploited and this can lead to the spontaneous support of the punishment organization and to emergence of some kind of social normativity concerning pool punishment. In this context, key conditions supporting the spontaneous emergence of coordinated pool punishment are (i) the willingness of some community members to engage in costly altruistic punishment and (ii) the possibility for altruistic punishment/sanctioning to become a social norm.

This can best be considered as follows: thanks to punishment mechanisms, a minority of reciprocal subjects effectively induces a majority of selfish subjects to cooperate.

3.2.3 Behavioral Dispositions: Trust

Trust, namely, a grounded belief about other individuals' positive attitudes, is a fundamental behavioral disposition promoting cooperation in social interaction.

The disposition to trust other agents may be grounded on our either direct or indirect acquaintance with their past behavior or on a rational calculus about their short/long-term interests.

Interestingly, there may be cases in which we lack any acquaintance with individuals, groups, or organizations, but it is still in our interest to enter profitable (commercial or political) reciprocation with them. In these cases (Bicchieri et al. 2004), individuals' expectations concerning other parties' trustworthiness may be adaptive, meaning that they are built on learned information about most fitting behavioral patterns, also gleaned from anonymous interactions.

The notion of trust has been importantly investigated in decision theory in the context of the so-called trust game (Berg et al. 1995), which is a variation of the dictator game, both of which are designed to allow the emergence of mutual confidence in strategic interactions. In this game, trust is encouraged by supplying minimal information about the other player's disposition, e.g., the willingness to allocate a high/low percentage of a received gift on behalf of the partner. Clearly, in this game, possible predictions or trusting expectations involve minimal knowledge of social contexts and other players' behavioral attitudes (Ostrom and Walker 2003).

Interestingly, it is often maintained that knowledge leading to trusting expectations can be acquired by means of so-called social inference: individuals can consider possible interactive roles of other players as expected sequences of behaviors and recall such sequences of expectations based on situational cues. Similar cognitive constructs, also defined as "schemata" (Bicchieri 2006), are formed on the basis of observations of repeated behavior or other forms of learning, stored in memory, and often shared, becoming common knowledge.

But there is a particular aspect of trust that makes it different from any other kind of expectation concerning the likely behavior of others (e.g., for prediction or control purposes), and this is the fact that trust is a matter of interest (Gambetta 1988, p. 222). Importantly, trust is a disposition to engage in social exchanges which are uncertain but also potentially rewarding (Bicchieri et al. 2004). Trust can be therefore defined as an interest-based belief: we are interested in the belief that other players will be cooperative, because the fact of cooperation itself will serve our interest.

It is therefore worth considering that, as trust is the belief through which cooperation can be predicted, it is often grounded not only on objective information, but also on a subjective estimation of risk. Trust can be thought of as crossing the personal threshold of risk acceptance/avoidance which triggers an individual's engagement in a cooperative endeavor (Gambetta 1988).

Depending on our background information and framing attitudes (Kahneman and Tversky 1979), we may or may not trust the probability that effective cooperation occurs in a social context (Bicchieri et al. 2004; 2006). Even rationally motivated cooperation may not emerge, simply because people don't trust each other enough to act on those motives (Gambetta 1988). Game theory, after all, has provided examples where cooperation in strategic interaction fails to take place even when it is rationally consistent to behave cooperatively (Binmore and Dasgupta 1986).

On the other hand, it is also worth considering that being interested in reciprocation as a way to promote coordination/collaboration means being interested in promoting mutual trust. Reciprocity mechanisms intersect with individuals' intentions to promote mutual trust: signaling, reputation building, and learned information about other individuals' reputations become important instruments in this regard (Gambetta 1988). Philosophers of social science (Pettit 1995) also focus on self-enhancing dynamics of mutual trust, so-called trust mechanisms, based on the fact that manifesting one's trust in someone can motivate that person to do what one is trusting them to do.

Overall, individuals' trust in interactions is a latent social asset. Societies relying on punishment and sanctioning systems are more stable but less efficient, more costly, and more unpleasant than those in which trust is maintained as an incentive to spontaneously engage in cooperation (Gambetta 1988). Also, individual and inter-organization relationships within the economic and political scenarios (especially in the frame of international conflict, Schelling 1966) benefit from trust-based reciprocity. Trust is a "social lubrificant" (Arrow 1974), especially in areas where transactions are dominated by incomplete contracts. In this regard experiments on the enforcement of non-binding agreements (Fehr and Gächter 2000; Fehr et al. 2002) suggested that trusting beliefs deriving from reciprocity mechanisms are a potent substitute for law when compliance with contracts is not explicitly regulated and therefore imperfectly enforced.

Nonetheless, trust is a vital and at the same time fragile commodity: lacking any binding character by definition, trust cannot be a stable behavioral trait. Mutual trust may degenerate into mutual distrust unless higher-order regulation intervenes: e.g., in the case of punishment, a central authority (institutional subject), assuming long-term obligation costs, can mitigate instability problems.

3.3 The Emergence of Institutions

Sociologists identify institutions as endurable regulators of human actions, usually characterized by specific roles and power relationships, that organize and structure social life at different levels of, from communities to markets and governments (Giddens 1984). Social scientists for the most part identify institutions with social norms, basically considering them shared normative patterns endowed with repeatability criteria and functioning as coordination devices.

A main distinction concerning this notion (North 1991) is that between formal and informal institutions. While formal institutions are rule-based social organizations which are legally designed, informal institutions include the variety of behavioral codes characterizing the spontaneous structure of a group: ethical rules, rituals, social conventions, and so forth. Informal institutions can be the result of voluntary and robust group organizations and are often able to solve common pool problems without relying on formal coercion (Ostrom 2005).

A main feature characterizing institutions in current views (Bicchieri 2006) is that they are all human artifacts, precisely, mind-dependent artifacts or social "constructions," primarily existing thanks to individual cognitive representations and actions (in contrast with the structure of physical reality), which are in turn able to condition individual representations and actions themselves.

The bottom-up emergence of artifacts (i.e. the emergence of institutions, which derive from individuals' voluntary attitudes and not from statutory laws) is apparently puzzling, because voluntary compliance with social standards implying personal costs apparently runs against evolutionary stability of individuals.

Game-theoretic approaches generally assume folks theorems concerning the long-term maximization of payoffs based on individual utility functions, in order to justify compliance with cooperative standards in social interactions (Bicchieri 2006; Binmore 2010). In this view standards for equilibrium are voluntarily or non-voluntarily reached by agents and then reinterpreted as institutions, including possible forms of enforcement. As mentioned earlier, possible justifications for other-regarding preferences in one-shot interactions (so-called social preferences), contrasting with the traditional assumption of self-regarding rationality in classical economic and rational-choice theory, have been linked with considerations of group benefit, based on the idea that members of a group benefit from mutual adherence to behavioral standards (Bowles and Gintis 2003). Parallel accounts emerged in a raging theoretical debate about the ontology of social preferences and the collective-choice process (Sugden 2015). These arguments exceed the scope of present considerations.

Interestingly, recent theories on the emergence of social artifacts focus on both motivational and cognitive traits of individual social-norm compliance, in search of evidence that contextual factors prime social conformity. So-called conditional preferences (Bicchieri 2006) and factors of internalization (Conte et al. 2014) have both been considered as emerging in relation to salience of social norms, i.e., to the perceived tendency of other individuals' compliance to behavioral standards, above a certain degree.

Particularly, conditional preferences (Bicchieri 2006; 2010) reflect the logic of reciprocity: individual preferences for complying with a social convention/norm occur "on condition" of positive expectations (empirical or normative) about others' compliance: I cooperate if you do also. In this context a social norm's stability is a function of the stability of the expectations that support it. While empirical expectations map what an individual expects other individuals to do in contingent strategic interactions, normative expectations are based on other people's expected behaviors, i.e., what other people believe they "ought" to do in certain situation. The degree of normative expectations (which can be grounded on past observation or indirect knowledge or even the projection of conformity) in a sense reflects the degree of salience of the supported social norms.

Accounts of conditional preferences for social conformity (Schelling 1966; Bicchieri 2006) generally are endowed with robust explanatory power in the investigation of individual attitudes toward social institutions. They usually assume such conditional preferences to be ultimately anchored to individual utility functions (in a game-theoretical perspective) and especially to be context dependent (Paternotte and Grose 2012); i.e., conditional preferences may change, also in one-shot interactions, according to variations in social contexts (such as variations of relations and roles).

It is worth considering in this regard that unconditional preferences for norm compliance entail a certain explanatory potential in order to account for the emergence and (especially) maintenance of social structures; nonetheless, any reference to social conformity as consistent with individual unconditional preferences and not well-defined personal values risks resulting in a simplistic view, failing to provide a correct focus on the social arena, while illegitimately trespassing in the analytical domain of morality. In contrast, the social arena remains the space within which normative artifacts emerge, propagate, and die out, also in contrast with self-interest motives and even in conflict with widely recognized moral principles (e.g., health protection). Social contexts are therefore the place where conditional determinants for social-norm compliance are to be searched.

3.3.1 Perspectives on Social Conditionality

Notwithstanding that the assumption of manifest payoff maximization was discarded in about 1990, the empty notion of utility function has been preserved in most game- theoretical approaches, commonly relying on rational-choice perspective. The general character of utility functions, made to fit any possible explanation of unobservable motivational determinants of behavior, tells us very little about individual preference mechanisms within social interaction. Still, this standard weakness helps maintain an important theoretical perspective: i.e., that utility preference (e.g., preference for benefit or welfare) and preference for action (e.g., voluntary compliance) are rationally consistent each other; the former rationally induces the latter.

As dispositions to cooperate are still cost-sensitive, we must assume that effects of cooperation represent a benefit (b) greater than any cost for attainment (c). Individual engagements in social compliance must rely on the assumption that, in order to gain or maintain a benefit x, it is worth a shared (even costly) compliance to y. Equally, individual engagement in reciprocation plausibly rests on a similar assumption, very simply expressed by the formula *if preference X, then preference Y.*

As interesting as the theoretical implications of this perspective can be, it is important to focus on preferences for social-norm compliance in terms of preferences for conformity—namely, we promote the view that a shared regularity is preferred (and therefore followed) under various conditions of sociality, while it stops being preferred if these conditions are no longer met. Unlike moral norms, which unconditionally induce compliance to behavioral patterns no matter what, social norms are complied with conditionally.

This being the case, the following discussion won't focus on the motivational factors inducing individuals to prefer to engage in a particular norm or behavioral route, which may also include incentives and sanctions, besides specific or correlate normative directions. Rather, we will briefly focus on the factors, which make a norm a social norm and on some of the factors responsible for its emergence, propagation, and resilience.

Overall, there is much evidence that people comply with social norms even in the absence of any manifest incentive structure or personal commitment to what the norm stands for (Cialdini et al. 1990). Indeed, assuming a conditional preference for following a fairness norm is different from assuming a fairness preference. As mentioned, game-theoretic accounts of social artifacts (Bicchieri 2006; Binmore 2010) make reference to conditional preferences for social- norm compliance as triggered by context-dependent factors and also present formal models of incorporation of social norms into individual preferences, which are based on knowledge of social contexts (cf. the notion of norm-based utility function in Bicchieri 2006, p. 115).

Basing on such theoretical evidence we draw attention to: (i) the fact that the representational content of a social norm, namely, the normative pattern suggested by a social norm, can be distinct from the motivations for which a similar norm is to be followed (and eventually enforced); (ii) the fact that such individual motivations for social-norm compliance, being grounded on individual utility, are basically conditional to social contexts.

A plausible explanation for the question why conditional preferences do emerge in social contexts (Bicchieri 2006; Paternotte and Grose 2012) is that, in a sense, they express both the requisite and the reason for a social norm to be followed and enforced: I conform "if" other group members also conform, but also I conform "because" other group members conform.

This apparently trivial argument is indeed the very condition for individuals' convergence on shared behavioral/normative routes (i.e., for cooperation) and also reciprocity mechanisms. We can plausibly assume that others' fairness preferences may influence individuals' behavior in repeated interactions in a way that renders selfish calculus misleading or inconsistent (direct reciprocity). Also we assume that if there emerges the awareness that there are to be continuing practices of conformity to, for example, norms of fairness in a social context, then people belonging to that social context must be motivated, one way or another, to enter "the rules of the game" (indirect reciprocity). Even the behavior resulting from an internalized disposition to contribute to a cooperative endeavor is contingent upon cooperation of others, on the condition of voluntary costly punishment (strong reciprocity).

Experimental evidence on cross-cultural variations of fairness (Henrich et al. 2006, 2010) show that the extent to which fairness norms are internalized depends on social contexts variables. The dictator-game experiments showed that the more involved participants were with markets, the more they tended to "give."

Reciprocation appears therefore to be a basic condition for conformity to social patterns (whatever their normative content), assuming that the logic driving individual motivational attitudes is a logic of utility preferences: I prefer (benefit from) reciprocation, then I prefer to share a behavioral or normative pattern. Put otherwise, representations of social constructs are anchored in individual preferences as functions of reciprocation, assuming that by isolating the representation of social patterns from the condition to which an individual's motivational attitude for social conformity is to be anchored, it is possible to regard social behavioral or normative patterns as placeholders of such functions.

A similar perspective, concerning conditional preferences for norm compliance as basic functions of reciprocation, plausibly grounds any evolving relationship between an individual and his family and, later, his society. Precisely with respect to different kinds of reciprocal relationships, it is possible to assume an evolving ordering of conditional preferences that is able to influence priorities of compliance to different kinds of social routes, from informal to formal institutions (Gerace 2013a). Arguments of this kind are beyond the scope of this discussion.

On the other hand, it is equally possible to consider the emergence of reciprocity functions as those concerning any behavioral standard directly or indirectly concerning the boosting and maintenance of conditions of reciprocity (i.e. society). Reputation, punishment, and trust are examples of pre-institutional functions of reciprocity: if reciprocity/society matters, then functions of reciprocity matter as social assets.

3.3.2 Models of Emergence

The idea that conditional preferences for social conformity basically rely on functions of reciprocation (and possibly of conditions of reciprocity) can help us to understand some aspects of the problem of emergent properties of social artifacts.

For more than a hundred years, the most challenging problem concerning the bottom-up emergence of social artifacts was to account for such complicated normative structures, which are difficult to concert, design, and even execute (Hume 1888, p. 538) but which are still able to engage a considerable number of individuals in uniform directions.

The challenge becomes even greater in considering evidence that most such structures emerge thanks to individuals' indirect coordination and self-enhancing propagation of standards. A further important challenge therefore concerns a possible account for the endogenous robustness of informal social artifacts emerging from individuals' interactions without planning or design.

Evolutionary game theory (Bowles and Gintis 2011) has mainly endorsed the view that the unintended emergence of regular patterns of coordination is due to processes of mutual adaptation, imitation, and cultural transmission, integrating a progressive self-enforcement of regularities thanks to the interplay between internalization processes of resulting norms themselves and norms of punishment.

On the other hand, economists and social scientists (Alexander et al. 1987; Goldstein 1999) have addressed the concept of the social system, focusing on complexity relations between individual actions and overall societal behavior, i.e., on the interplay between lower- and higher-level variables of social interaction (so-called micro–macro dynamics), while especially investigating the status and resilience of the latter. Emergence in complex systems like societies is conceived as a process whereby higher-order entities (behavioral patterns or regularities) arise from the self-organizing interaction among lower-order entities (agents), without latter's exhibiting any of the properties of higher-order entities: the higher order is irreducible to the former, while being able to influence it.

In recent accounts (Conte et al. 2014) social scientists have considered that complexity dynamics concerning the emergence of social artifacts are not to be confined to the traditional view (Goldstein 1999) that radical novel macro-effects are generated by micro-dynamics in agents' interaction (simple loop). The non-deliberative, unperceived production of social phenomena is instead connected to the two-way process of emergence–immergence, also defined as a "complex recursive loop" (Conte et al. 2014, pp. 23, 46).

In the simple loop, a lower-level system produces an emergent effect at a higher level (e.g., the phenomenon of reputation emerges from repeated interactions); the emergent effect retro-acts on the lower level by determining a new property of the generating system (e.g., reputation affects agents' interactions by means of new behavioral traits, namely, more-or-less trusting relationships). In the complex loop, instead, the emergent effect is able to determine new properties at the lower level, by means of which the same effect is reproduced again, mostly in a stable way and with global significance. Emergent effects are likely to be reproduced thanks to so-called incorporation of new properties, which is a more-or-less conscious process directly responsible for the modification of mechanisms through which lower-level entities operate. Precisely, incorporation can occur as (i) second-order emergence, i.e., by means of recognition and aware reproduction/support of the emergent effect; and (ii) immergence, i.e., by means of the non-deliberate changing of mechanisms governing agents' decisions and interaction (e.g., unconscious internalization of social normativity), with consequent unaware reproduction of the effect.

Here is a possible example of second-order emergence in social interaction: reputation is an emergent phenomenon. Beliefs about the positive function of reputation and consequent adoption of relating attitudes, such as reputation building and image scoring, reproduce the phenomenon with positive global effects: maintenance of cooperation stability through indirect reciprocity.

On the other hand, innumerable examples of immergence and unaware reproduction of social effects, unfortunately also negative ones, take place right under our noses, e.g., in financial markets or in the political scenario.

Drawing on an evolutionary view, the emergence–immergence perspective focuses on adaptive mechanisms and background conditions of internalization. Interestingly, immergence is regarded as the precondition for individual motivations to social conformity, to reach indirect coordination in a way similar to the swarming behavior of lower species: stigmery (Conte et al. 2014, pp. 23). It is considered possible that individual subsequent actions tend to reinforce and build on each other, with no need for direct communication to take place in order to allow for non-deliberate clustering, conformity, or coordination effects. In this context salience is regarded as both an incentive and an effect of immergence processes.

It is worth a question then: which mechanisms can be affected by the immergence process, in a way that shapes such motivational intelligence in the social arena? A plausible answer could be preference mechanisms. The interlocking of preferences from different agents, also unknown to each other, can determine an indirect, mostly non-represented convergence of individual motivational attitudes toward social conformity and the consequent non-deliberate building of social structures. In this context social artifacts can shape as a byproduct of actions directed toward different kinds of ends.

Yet, as we mentioned, even assuming a variety of personal preferences and utility functions as the ultimate reason for social conformity, the effective preconditions and therefore the determinant variables for voluntary engagement in contingent long-term "shared" obligations seems to be better found in preferences for reciprocity and sociality[1]. Individual preferences for sociality can be regarded as the anchor of shared behavioral/normative patterns.

On the other hand, assuming that social patterns nest in each other in such a way as to produce complex normative structures, it is possible to understand how social artifacts can emerge and be efficiently supported even against contingent self-interest and especially without strong informational background about their design, planning, or control (Gerace 2012).

Unfortunately, the resilience of emerging social artifacts is not an absolute one. As mentioned earlier, group size strongly influences the degree of group cooperation (Carpenter et al. 2009). Experimental evidence showed that emerging social effects can reproduce almost constantly under certain conditions, but environmental variables such as size of population, demographic distribution, and frequency of interactions can strongly influence their organization and evolution.

It is therefore maintained so far that emerging institutions are mostly unstable structures, in need of centralized coordination able to ensure long-term cooperative agreement (independent from dynamic variables of interaction), on pain of collective action failure and negative consequences on the global social outcome.

In the following section we will present an experimental study aimed at demonstrating how emergent artifacts (and possible instability) don't only concern specific behavioral patterns (such as courtesy, for instance) or explicit normative directions (such as norms of fairness and such), but also latent behavioral attitudes eventually boosting cooperation, such as collaborative tendencies and trusting dispositions.

3.4 An Experiment in Social Interaction: The Public Trust Model

Previous studies (Bicchieri et al. 2004) focused on trustworthiness as behavioral regularity in bounded rationality conditions, showing how an evolutionarily stable state dominated by behavioral patterns of trusting and reciprocating (i.e., by a social norm of trust) may emerge in a repeated trust game, where strategies are not just history-contingent but also role-contingent. Importantly, it has been showed that repeated patterns of generalized impersonal trust emerged as the outcome of several different conditional strategies (where players trusted/reciprocated on condition of reciprocation), irrespective of the specific methods used by these strategies to elicit reciprocity.

Here, we will instead present a preliminary investigation concerning the emergence of trusting behavior in terms of collaborative tendency, not as a behavioral

[1] Sanctions, for instance, can be rightly regarded as exogenous incentive for support of sociality, but even sanction efficacy must count on some preference for social reciprocation: if I could do without sociality, I would also take no interest in sanction.

regularity itself but rather as a gradually increasing attitude strongly influenced by the salience of cooperativeness, in a sort of social contagion. We will also examine whether such a trust climate is a self-enhancing stable social phenomenon capable of persisting despite changes of variables, or not (Gerace and Cecconi 2014).

3.4.1 Model Description

Our model relies on a random group design, precisely, a situation in which *N* variable agents have random possibilities to perform and repeat strategic interactions, according to two main parameters: (a) visual space of interaction and (b) velocity of interaction.

The kind of interaction we rely on is the donation game, namely, a variation of the iterated prisoner dilemma, where cooperation corresponds to offering the other player a benefit b at a cost c with b > c. Defection means offering nothing. According to the payoffs relationship of the donation game an iterated mutual cooperation, i.e., 2(b−c), is always the best strategy, so that each player needs to rely not only on the possibility to repeat interactions, but also on the possible opponent's disposition to cooperate in the next rounds.

In a highly simplified example the donation game shows how the mechanism of indirect reciprocity operates using both payoff-relevant information and individuals' reputations to promote cooperation. But, unlike the standard model of indirect reciprocity, which offers a binary choice—people can either cooperate or defect—in our model agents can increase or diminish their collaborative tendency in a range from 0 to 1. The degree of collaborative tendency is assumed to indicate the degree of trusting expectations about others' positive dispositions.

Basic assumptions of the model are that: (i) agents have an original behavioral disposition varying randomly from 1 (cooperative) to 0 (non-cooperative); (ii) agents can form expectations about others' behavioral tendencies; and (iii) agents tend to improve their status.

The main algorithms in the model work as follows: (i) agents have memory of past interactions and record collaborations; (ii) agents are influenced by a certain threshold of others' payoff target, in a way that imitates their cooperative disposition; and (iii) agents can perform interactions according to four independent simulation modes.

The first simulation mode, named "none," is a simple reciprocity-based model of cooperation, where agents play without recording collaborative encounters, increasing their collaborative attitude by imitating better-performing agents in their visual space.

The second mode, named "direct reciprocity," is an adaptive reciprocity-based model of collaborative behavior, where agents increase their cooperativeness (in other words, their trusting expectations) in contingent interactions, if they experienced enough reciprocation in their past experience: they trust their partner if the memory-rate of their past collaborative encounters exceeds a certain threshold, or defect. The third mode, named "individual trust," is a reputation-based mode of emergence of stable expectations, where agents progressively increase their general collaborative disposition, i.e., their general trusting expectations, based on the level of their past positive encounters.

The last mode, "public trust," is a salience-based interaction mode allowing the emergence of stable trusting expectations: agents increase their general collaborative tendency based on the general degree of cooperativeness or trusting behavior within the population.

The public-trust mode may offer a possible example of immergence effects. We can assume an emergence–immergence process leading to the bottom-up emergence of public trust as a social artifact: if cooperation becomes the dominant observed behavior (emergence), this triggers a positive, possibly subliminal, influence in individual trusting expectations (immergence), so that and a new artifact emerges, namely, public trust or a trust climate, where impersonal/general trust becomes a self-enhancing shared behavioral direction.

3.4.2 Results

The first four-modes simulation was run using the following parameters: number of agents: 500; visual space: 0.1; velocity of interaction: 1. The first results are shown in Fig. 3.1.

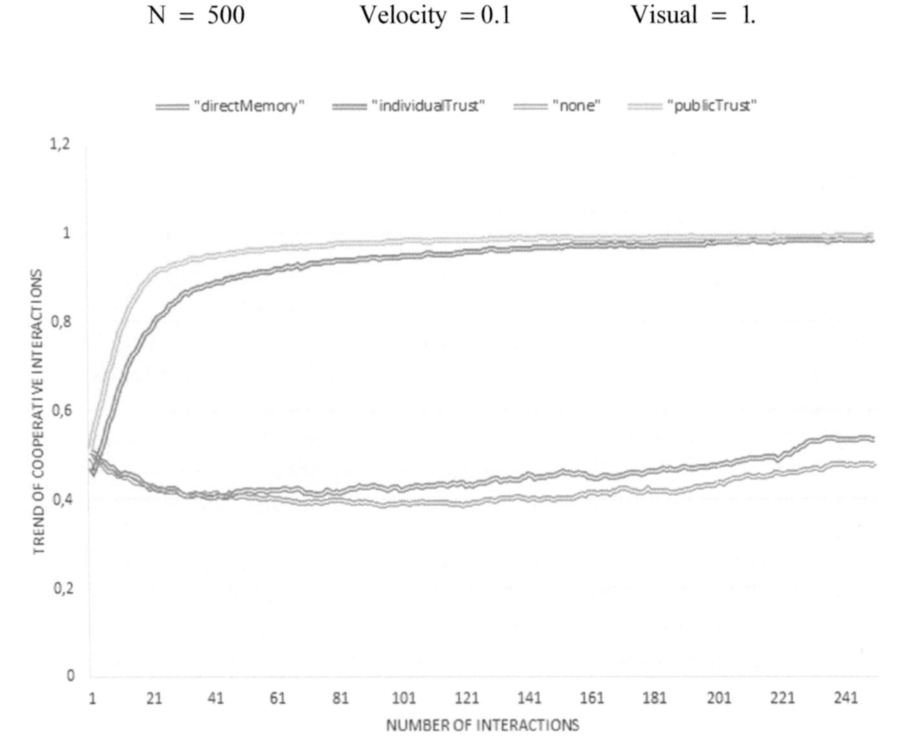

Fig. 3.1 Mean of agents' behavior

The figure shows that the trend of cooperative interactions is set at the highest levels of the range for the individual-trust mode and even more for the public-trust mode. In contrast, a slow increase of collaborative behaviors toward the middle of the range characterized the none mode and (to a slightly greater degree) the direct-memory mode. None of the interaction modes, however, exhibit a decreasing trend.

A second four-modes simulation was run on different values of visual space and velocity of interaction, respectively, 5 and 0.5. As shown in Fig. 3.2, here the most evident results were seen in the public-trust mode.

The public-trust trend collapses, unsteadily settling below middle levels, while both the direct-memory and none modes settle at minimum levels; only individual trust shows a steady growth of cooperation.

3.4.3 Discussion of Results

The most interesting aspect of these results is the clear sensitivity of higher levels to lower levels of interaction, in the absence of noise. In fact, the emergence of stable expectations in typical indirect-reciprocation modes (individual and public trust) is strongly influenced by preliminary conditions of cooperation.

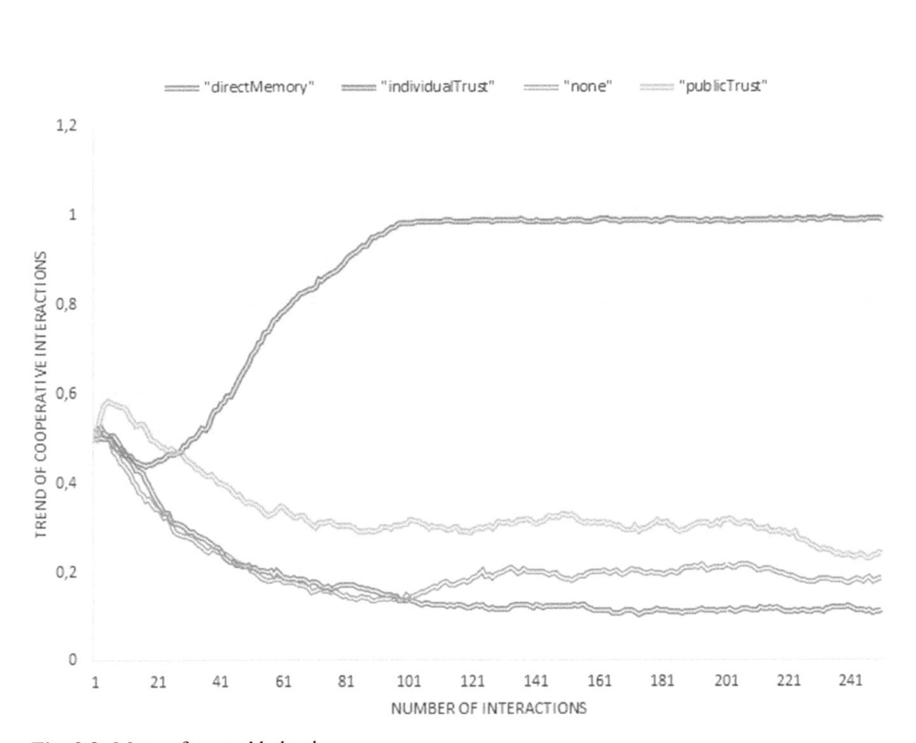

Fig. 3.2 Mean of agents' behavior

This is particularly true for public trust. As reflected by direct reciprocity results in Fig. 3.1, general cooperation (emergence and slow increase) is granted above a certain threshold thanks to favorable space–time conditions' allowing for a sufficient frequency of interaction and encounters iteration; clearly the same favorable conditions also allow for a sufficient degree of salience in the public trust mode. Conversely, as shown by Fig. 3.2, when environmental changes occur (such as higher space and lower velocity of interaction), this increases the probability of differentiating interactions while diminishing the possibility of iterating them, so that not only does direct reciprocity-based cooperation decrease, but so does the salience-based mode of cooperation, i.e., public trust.

Given the assumption that individuals engaging in both kind of interactions (direct memory and public trust) keep track of the scores of positive encounters (in their own experience or in their community respectively), then it has to be considered that the high probability of different interactions makes it difficult for individuals to update their image-scores. On the other hand, the payoff-maximization strategy of the donation game (triggering the self-enhancing chain of cooperation through the imitation of best performers) requires players to be matched with the same partners at least once.

A question arises concerning the reason that individual-trust mode doesn't show the same sensitivity to environmental changes as public trust. An intuitive reason for this is the fact that individual trust is not an emergent phenomenon in the sense of public trust. Agents of individual trust interaction mode don't need to rely on preliminary conditions of general cooperation, but only to their memory-rate, in order to progressively increase their collaborative disposition: this makes it possible for cooperation to emerge quite soon after the beginning of the play even in one-shot interactions. Average positive payoff is achieved, so that the self-enhancing process of reciprocation is then added to the process of imitation of best performers (which is interpreted as a learning process).

Three out of four simulated-interaction modes self-enhanced toward positive or negative directions on condition of reciprocity. Public trust in particular presented as an emergent social artifact that was non-derivative of any enforcement strategy, but only of conditional preferences, according to the following rule: always trust if the community is collaborative above a certain threshold.

On the other hand, public trust didn't present as a structurally stable phenomenon, evolving independently of conditions of interaction. This confirms the theories according to which emergent social artifacts require formal institutions in order to gain stability and permanent resilience.

3.5 Conclusions

We provided some evidence that cooperation, mainly regarded as the capacity to comply with long-term shared obligations, is correlated with individuals' willingness to establish (direct or indirect) reciprocal relations outside the domains of family and fellowship.

In order to provide theoretical justification for pro-social and costly motives, generally thought of as evolutionary effects of social learning and cultural transmission, we assumed the apparently trivial perspective that individual social behaviors are triggered by rational mechanisms of preferences and more exactly preferences for sociality. Much of what has been defined the anchoring of preferences for any shared norm-governed practice or shared behavioral tendency has been theoretically found in the condition of sociality (or reciprocity) itself.

We considered the possibility that a similar perspective fits some convincing arguments concerning the informal emergence of social phenomena, in terms of spontaneous convergence upon shared regularities, at least initially independent from formal enforcement solutions. This perspective particularly fits the so-called emergence–immergence process: individual preferences in micro-interactions induce macro-phenomena (social patterns of behavior), to which individuals themselves adapt with new preferences, automatically reproducing macro-effects (social conformity). In this context, individual preferences for conformity are assumed to be conditional to macro-phenomena in terms of salience of social patterns.

As most likely non-strategic and not forward-looking, there are behavioral tendencies for conformity to novel or well-known social patterns, which just in following the genuine rationality of preferences can build very articulated and more-or-less stable social structures (instead of relying on implausibly complex representational attitudes).

If a theory does not explain anything unless it points to underlying causal mechanisms, then an understanding of such rationality in individuals motivational attitudes seems to be the appropriate route to comprehend structures and forces lying behind social phenomena.

Yet, even assuming a plausible context dependency of individual preferences for social conformity (particularly through salience of behavioral patterns or dispositions), the specifics of how social contexts affect changes in individual preferences remain fairly vague (Paternotte and Grose 2012).

Most of what concerns the emotional and unreflective determinants of both social- norms priming and context-dependent compliance hasn't been treated with systematic accuracy thus far. The important need now is therefore a theory of how preferences emerge, dynamically change, and influence individual social engagement, at all levels of individuals' motivational attitudes (Gerace 2013b). On these bases decision theorists and social scientists, engaged in formal modeling, could also be provided with adequate conceptual resources for capturing the role played by preferences in social interaction.

References

Alexander, R. D. (1987). *The biology of moral systems*. New York: De Gruyter.
Alexander, J. C., Giesen, B., Munich, R., & Smelser, N. J. (Eds). (1987). *The micro-macro link*. Berkeley: University of California Press.
Arrow, K. J. (1974). *The limits of organization*. New York: W. W. Norton & Co.
Axelrod, R. (1984). *The evolution of cooperation*. New York: Basic Books.

Axelrod, R., & Dion, D. (1988). The further evolution of cooperation. *Science, 242,* 1385–1390.
Axelrod R., & Hamilton W. D. (1981). The evolution of cooperation. *Science, 211,* 1390–1396.
Berg, J., Dickhaut, J., & McCabe, K. (1995). Trust, reciprocity, and social history, games and economic behavior. *Games and Economic Behaviour, 10,* 122–142.
Berger, U., & Grüne, A. (2014). Evolutionary stability of indirect reciprocity by image scoring. Department of Economics Working Paper Series 168.
Bernhard, H., Fischbacher, U., & Fehr, E. (2006). Parochial altruism in humans. *Nature, 442,* 912–915.
Bicchieri, C. (2006). *The grammar of society*. Cambridge: Cambridge University Press
Bicchieri, C. (2010). Norms, preferences and conditional behaviour. *Politics, Philosophy, and Economics, 9,* 297–313.
Bicchieri, C., Duffy J., & Tolle G. (2004). Trust among strangers. *Philosophy of Science, 71*(3), 286–319.
Binmore, K. (2010). Social norms or social preferences? *Mind and Society, 9,* 139–158.
Binmore, K., & Dasgupta, P. (1986). Game theory: A survey. In K. Binmore & P. Dasgupta (Eds.), *Economic organizations as games*. Oxford: Basil Blackwell.
Bowles, S., & Gintis, H. (2004). The evolution of strong reciprocity: Cooperation in heterogeneous populations. *Theoretical Population Biology, 65*(1), 17–28.
Bowles, S., & Gintis, H. (2011). *A cooperative species: Human reciprocity and its evolution.* Princeton: Princeton University Press.
Boyd, R., Gintis, H., & Bowles, S. (2010). Coordinated punishment of defectors sustains cooperation and can proliferate when rare. *Science, 328,* 617–620.
Brandt, H., Ohtsuki, H., Iwasa, Y., & Sigmund, K. (2006). A survey on indirect reciprocity. IIASA Interim Report, IR-06-065.
Carpenter, J., Bowles, S., Gintis, H., Ha-Hwang, S. (2009). Strong reciprocity and team production: Theory and evidence. *Journal of Economic Behavior & Organization, 71*(2), 221–232.
Cialdini, R. B. (1993). *Influence: Science and practice* (3rd ed.). New York: Harper Collins.
Cialdini, R. B., Kallgren, C. A., & Reno, R. R. (1990). A focus theory of normative conduct: A theoretical refinement and reevaluation of the role of norms in human behavior. *Advances in Experimental Social Psychology, 24,* 201–234.
Conte, R., Andrighetto, G., & Campennì, M. (Eds.). (2014). *Minding norms. Mechanisms and dynamics of social order in agent societies*. Oxford: Oxford University Press.
De Waal, F. B. M. (1996). *Good natured: The origins of right and wrong in humans and other animals*. Cambridge: Harvard University Press.
Fehr, E., & Gächter, S. (2000). Cooperation and punishment. *American Economic Review, 90,* 980–994.
Fehr E., & Gächter, S. (2002). Altruistic punishment in humans. *Nature, 415,* 137–140.
Fehr E., & Fischenbar, U. (2003). The nature of human altruism. *Nature, 425,* 785–791.
Fehr, E., & Schmidt, K. M. (1999). A theory of fairness, competition, and cooperation. *The Quarterly Journal of Economics, 114*(3), 817–868.
Fehr E., Fischbacher U., & Gächter S. (2002). Strong reciprocity, human cooperation, and the enforcement of social norms. *Human Nature, 13,* 1–25.
Gächter, S., & Herrmann, B. (2009). Reciprocity, culture and human cooperation: Previous insights and a new cross-cultural experiment. *Philosophical Transactions of the Royal Society: Biological Sciences, 364,* 791–806.
Gambetta, D. (1988). Can we trust trust? In D. Gambetta (Ed.), *Trust: Making and breaking cooperative relations* (pp. 213–237). Oxford: Basil Blackwell.
Gerace, G. (2012). Justifying normative objects: An answer to J. Searle's model. Paper presented at the 20th Conference of the Italian Society for Analytic Philosophy (SIFA), September 12–15, Alghero, University of Sassari.
Gerace, G. (2013a). Emergence and persistence of social norms: A material-based account. Paper presented at the Mid-term Postgraduate Conference of the Italian Society of Logic and Philosophy of Science (SILFS), May 29–31, University of Urbino.

Gerace, G. (2013b). Justifying Normative Objects. Project proposal awarded a 2013–2017 Phd position in Philosophy of the Social Sciences—College of Social Sciences and International Studies, University of Exeter (UK).

Gerace, G., & Cecconi, F. (2014). The emergence of collaborative behaviour. Paper presented at *Carving Societies—workshop*, 1 December 2014, LABSS-ISTC-CNR, Rome.

Giddens, A. (1984). *The constitution of society: Outline of the theory of structuration*. Cambridge: Polity Press.

Gintis, H. (2000). Strong reciprocity and human sociality. *Journal of Theoretical Biology, 206*(2), 169–179.

Gintis, H. (2004). The genetic side of gene-culture coevolution: Internalization of norms and pro-social emotions. *Journal of Economic Behaviour and Organization, 53,* 57–67.

Gintis, H., & Fehr, E. (2012). The social structure of cooperation and punishment. *Behavioral and Brain Sciences, 35*(1), 28–29.

Gintis, H., Alden S. E., & Bowles, S. (2001). Costly signalling and cooperation. *Journal of Theoretical Biology, 213,* 103–119.

Gintis, H., Fehr, E., Bowles S., & Boyd, R. (2005). *Moral sentiments and material interests: The foundations of cooperation in economic life*. Cambridge: MIT Press.

Goldstein, J. (1999). Emergence as a construct: History and issues. *Emergence, 1*(1), 49–72.

Gouldner, A. W. (1960). The norm of reciprocity: A preliminary statement. *American Sociological Review, 25,* 161–178.

Hamilton, W. D. (1964). The genetical evolution of social behaviour. *Journal of Theoretical Biology, 7* (1), 1–16.

Hauert, C., & Szabó, G. (2005). Game theory and physics. *American Journal of Physics, 73*(5), 405–414.

Hume, D. (1888). *Treatise of human nature*. Oxford: Clarendon Press.

Henrich J., et al. (2001). In search of homo economicus—Behavioral experiments in 15 small-scale societies. *American Economic Review, 91*, 73–79.

Henrich, J., et al. (2006). Costly punishment across human societies. *Science, 312*(5781), 1767–1770.

Henrich, J., et al. (2010). Markets, religion, community size, and the evolution of fairness and punishment. *Science, 327,* 1480–1484.

Kahneman, D., & Tversky, A. (1979). Prospect theory: An analysis of decisions under risk. *Econometrica, 47*(2), 263–291.

Kandori, M. (1992). Social norms and community enforcement. *The Review of Economic Studies, 59*(1), 63–80.

Milinski, M., Semmann, D., Bakker, T. C. M., & Krambeck, H. J. (2001). Cooperation through indirect reciprocity: Image scoring or standing strategy? *Proceedings of the Royal Society of London, Series B, 268,* 2495–2501.

North, D. C. (1991). Institutions. *The Journal of Economic Perspectives, 5*(1), 97–112.

Nowak, M. A., & Roch, S. (2006). Upstream reciprocity and the evolution of gratitude. *Proceedings of the Royal Society B: Biological Sciences, 274*(1610), 605–610.

Nowak, M. A., & Sigmund, K. (1998). Evolution of indirect reciprocity by image scoring. *Nature, 393*(6685), 573–577.

Nowak, M. A., & Sigmund, K. (2005). Evolution of indirect reciprocity. *Nature, 437,* 1291–1298.

Nowak, M. A., Tarnita, C. E., & Antal T. (2010). Evolutionary dynamics in structured populations. *Philosophical Transactions of the Royal Society of London, Series B: Biological Sciences, 365*(1537), 19–30.

Ohtsuki, H., & Iwasa, Y. (2006). The leading eight: Social norms that can maintain cooperation by indirect reciprocity. *Journal of Theoretical Biology, 239,* 435–444.

Ohtsuki, H., Iwasa, Y., & Nowak, M. A. (2009). Indirect reciprocity provides only a narrow margin of efficiency for costly punishment. *Nature, 457,* 79–82.

Ostrom, E. (2005). *Understanding institutional diversity*. Princeton: Princeton University Press.

Ostrom, E., & Walker, J. (2003). *Trust and reciprocity: Interdisciplinary lessons from experimental research*. New York: Russell Sage Foundation.

Panchanathan, K., & Boyd, R. (2003). A tale of two defectors: The importance of standing for evolution of indirect reciprocity. *Journal of Theoretical Biology, 224,* 115–126.
Paternotte, C., & Grose, J. (2012). Social norms and game theory: Harmony or discord. *British Journal for the Philosophy of Science*. doi:10.1093/bjps/axs024.
Pettit, P. (1995). The cunning of trust. *Philosophy and Public Affairs, 24,* 202–225.
Rand, D. G., & Nowak, M. A. (2013). Human cooperation. *Trends in Cognitive Sciences, 17*(8), 413–425.
Rand, D. G., et al. (2012). Spontaneous giving and calculated greed. *Nature, 489,* 427–430.
Rezaei, G., & Kirley, M. (2012). Dynamic social networks facilitate cooperation in the n-player prisoner's dilemma. *Physica A: Statistical Mechanics and its Applications, 391*(23), 6199–6211.
Rezaei, G., Kirley, M., & Pfau, J. (2009). Evolving cooperation in the n-player prisoner's dilemma: A social network model. *Artificial Life: Borrowing from Biology, 5865*, 43–52.
Rockenbach, B., & Milinski, M. (2006). The efficient interaction of indirect reciprocity and costly punishment. *Nature, 444,* 718–723.
Schelling, T. C. (1966). *The strategy of conflict*. New York: Oxford University Press.
Sigmund, K., & Nowak M. A. (2001). Evolution: Tides of tolerance. *Nature, 414,* 403–405.
Sigmund, K., De Silva, H., Traulsen, A., & Hauert, C. (2010). Social learning promotes institutions for governing the commons. *Nature, 466,* 861–863.
Sugden, R. (1986). *The economics of rights, co-operation, and welfare*. Oxford: Basil Blackwell.
Sugden, R. (1991). Rational choice: A survey of contributions from economics and philosophy. *Economic Journal, 101*(4), 751–785.
Sugden, R. (2015). Team reasoning and intentional cooperation for mutual benefit. *Journal of Social Ontology, 1*(1), 143–166.
Trivers, R. L. (1971). The evolution of reciprocal altruism. *Quarterly Review of Biology, 46*(1), 35–57.
Uchida, S., & Sigmund, K. (2010). The competition of assessment rules for indirect reciprocity. *Journal of Theoretical Biology, 263*(1), 13–19.
Wedekind, C., & Milinski, M. (2000). Cooperation through image scoring in humans. *Science, 288,* 850–852.
Wilson, D. S. (1975). A theory of group selection. *Proceedings of the National Academy of Sciences of the United States of America, 72,* 143–146.
Zahavi, A., & Zahavi, A. (1997). *The handicap principle: A missing piece of Darwin's puzzle*. New York: Oxford University Press.
Zahng, B., Li, C., Silva, H., Bednarik, P., & Sigmund, K. (2014). The evolution of sanctioning institutions: An experimental approach to the social contract. *Experimental Economics, 17*(2), 285–303.

Chapter 4
Modelling Extortion Racket Systems: Preliminary Results

Luis G. Nardin, Giulia Andrighetto, Áron Székely and Rosaria Conte

4.1 Introduction

Mafias may be defined as criminal organisations that are in the business of producing, promoting, and selling protection (Gambetta 1993).[1] They are widespread and can be found across the globe; the Russian Mafia is one incarnation that primarily operates in Russia (Varese 1996, 2001), the Yakuza in Japan (Hill 2006), and the Triads in Hong Kong (Morgan 1960). In Italy alone, there are three large and well-established mafias: the Sicilian Mafia in Sicily, the 'Ndrangheta in Calabria, and the Camorra in Campania (Savona 2012).

These criminal organisations cause both economic and social damage to the societies in which they are embedded (Daniele 2009). One reason is because they do not only offer their services to people and businesses that participate in legal transactions, but also—and perhaps more so—to those who are involved in illegal transactions, allowing markets for these illegal, and frequently harmful, goods and services, to exist (Gambetta 1993, pp. 226–244). They can also enforce cartels among businesses, driving up costs, hurting consumers, and reducing productivity (Gambetta 1993, pp. 195–225; Varese 2013, p. 5). One study estimates that the ma-

[1] The protection that mafias provide ranges over a continuum from the 'protection' from harm that the extorter would cause, to genuine protection—nevertheless socially harmful—that enforces cartels.

L. G. Nardin (✉) · G. Andrighetto · Á. Székely · R. Conte
LABSS. Laboratory of Agent-Based Social Simulation, Institute of Cognitive Sciences and Technologies—CNR, Via Palestro 32, 00185 Rome, Italy
e-mail: gnardin@gmail.com

G. Andrighetto
e-mail: giulia.andrighetto@istc.cnr.it

Á. Székely
e-mail: aron.szekely@istc.cnr.it

R. Conte
e-mail: rosaria.conte@istc.cnr.it

F. Cecconi (ed.), *New Frontiers in the Study of Social Phenomena*,
DOI 10.1007/978-3-319-23938-5_4

fias in Italy combined produce tax-free capital that was equivalent to approximately 7% of the national GDP in 2007 (Barone and Narciso 2013). Other studies have examined the economic harm caused by the Italian mafias, and at least two have found that the presence of mafias substantially hampers economic growth (Lavezzi 2008; Pinotti 2012).

Thus, overcoming, or at least limiting, the strength and influence of mafias is a societally beneficial objective. Yet, this is a difficult task, since the mafias that survive are the ones that are deeply entrenched in the societies within which they operate, often benefiting from the support of significant portions of society. This support may be based on a two main factors. First, they may provide some degree of genuine protection. Second, they employ their disproportionate power to intimidate and threaten, implicitly or explicitly, those who do not comply.[2]

One step to take towards defeating mafias is to deepen our understanding of them. The *Palermo Scenario* is an agent-based simulation model that contributes towards this objective. It can be used to address several research questions that are important to both policy-makers and researchers in understanding mafias and evaluating methods for destabilising them. It allows us to explore the independent and combined effects that different input variables, and actions by actors, have on destabilising a mafia. The model also enables the entire pathway to be investigated, not only from actions to mafia destabilisation, but also the intermediate actions along the path and actors' internal mental representations that favour their promotion.

It is reasonable to expect that a successful anti-mafia strategy should consider both the direct fight against them and an indirect approach that works on promoting socially beneficial behaviour among the population (i.e., the promotion of the culture of legality). An important element of the systems within which mafias operate is the interplay between the legal norms and the social norms. Legal norms or laws are rules of social behaviour that are established by a legal authority and enforced by specific third-party enforcers using legal sanctions (Elster 2007, p. 357). In contrast, social norms are socially shared rules or principles that prescribe what individuals should or should not do and that are often enforced through social sanctions. They are not as simple rules that individuals unconditionally comply with, but are shared beliefs and prescriptions regarding appropriate and expected behaviour in specific circumstances (Bicchieri 2006; Elster 2009; Conte et al. 2014).

Some specific research questions that can be answered by using our model are:

1. What effects do different policies of the state have on destabilising mafias?
2. How do independent and combined actions of different actors (such as the state, non-governmental organisations, and civil society) affect mafias?
3. Which conditions favour the spread of the culture of legality that undermine mafias?

Our core proposal is that that anti-mafia legislation, or specific legal norms, must be supported by social norms among the population within which a mafia functions to be effective at curtailing undesirable behaviour: there should be an alignment between

[2] Part of this is likely due to the selection effect for mafias in which those not entrenched in their milieu do not survive.

social and legal norms, while conflict between them leads to failure or only partial success. If this is the case, then it becomes crucial to establish the strategies that are most successful in promoting the abandonment of harmful social norms and how new beneficial social norms may effectively spread. Consequently, a link with the abundant theoretical and empirical literature on social norms can be made, and used to leverage our understanding of mafias. For new social norms to be adopted, actors need to change their beliefs, goals, and expectations and to be convinced that others have also changed their beliefs and will act accordingly. We posit, and test, that civil society organisations play an important role in coordinating the shift of actors' mental representations by public manifestations (by declarations, oaths or otherwise) and in favouring the spread of social norms supporting social desirable behaviour.

Our proposal has far reaching consequences, for it applies in important ways to the study of mafias, and their reduction. Yet, the same notion—congruence between social norms and legal norms, or more generally incentives, promotes behaviour change—applies more generally to other socially harmful practices including infibulation and female genital mutilation, foot binding, hand washing behaviour among doctors, binge drinking, and smoking.

The paper unfolds as follows. In Sect. 2, we describe the Palermo Scenario: our model that aims to represent the actors that are involved in the mafia phenomenon and the relationships among them. The Palermo Scenario will be used to check the research questions that we posed above. Next (in Sect. 3), we present an experiment that tests our predictions and that examines the independent and combined effects of different policies in countering mafias. In Sect. 4, we discuss the results that we have obtained so far. Finally, we provide some conclusions as well as some ideas for future work in Sect. 5.

4.2 Palermo Scenario

Based on iterative participators modelling and contemporary and historical empirical evidence extracted from a range of sources, we identified five key actors in the dynamics of the mafia phenomenon and their inter-relationships: Entrepreneurs, Consumers, the State, the Mafia, and an Intermediary Organisation.[3,4]

The model described here is a preliminary version of the Palermo Scenario that is under development and will be presented in its final form in future publications. The results we present are those obtained with this preliminary version of the model.

[3] These sources are judicial documents, confiscated Mafia documents such as *Libri Mastri* (accounting books used by some Mafiosi to record who various information about pizzo payers and that are occasionally discovered by the police), academic studies, literature, and other sources such as newspapers and television interviews.

[4] Twenty-seven expert stakeholders associated with the EU-funded GLODERS project participated at various time points between 2012 and 2015 in the model building process. See Nardin et al. (under review, p. 8) for details.

Entrepreneurs represent businessmen and liberal professionals. They are modelled as multiple agents and are the central actors in the model. They sell products to Consumers at a range of prices and receive income, and make a number of decisions using a combination of economic and normative reasoning. Entrepreneurs can

(i) Decide to pay pizzo[5] if approached by Mafiosi
(ii) Report pizzo requests to the State if they decide not to pay pizzo,
(iii) Report to the State damages that they sustained from Mafia attacks.
(iv) Collaborate with the State against specific Mafioso if approached by the State and finally, they can
(v) Join the Intermediary Organisation, thereby signalling that they are unwilling to pay pizzo, likely to report pizzo requests and Mafia punishments, and obtain respite from Mafia requests.

The State represents the Italian state. It can:

(i) Imprison Mafiosi. Mafiosi can be sent to prison after investigation by the police, who either work with specific evidence obtained from Entrepreneurs, or with evidence obtained from general day-to-day observation and police activity. Naturally, investigations based on specific evidence are more effective than those based on general observation. After the police captures a Mafioso, the police may find information about the Entrepreneurs who paid pizzo to that Mafioso: the Mafioso may provide information (i.e., pentiti) or the information may be found in assorted documents such as *Libro Mastro*. The State can then use this evidence to elicit collaboration from those Entrepreneurs by threatening them with punishment and if collaboration is obtained, the State uses their information to increase the possibility of prosecuting that Mafioso.
(ii) Support Entrepreneurs who have suffered damages at the hands of Mafiosi. Entrepreneurs who have suffered some damages from Mafia retaliation can apply for monetary support to a fund that is set-up specifically for this purpose, the *Fondo di Solidarietà* (i.e., a State-run fund to support Mafia victims), which contains resources that depend on a politically determined component and a component derived from the resources of captured Mafiosi.
(iii) Spreads facts about successful actions that it has carried out against the Mafia (consider this as the State providing information to journalists who report and propagate the news in newspapers and television programs) and it can work to
(iv) Change peoples' attitudes regarding the Mafia using campaigns and education regarding appropriate behaviour; some of which is done by sponsoring and supporting anti-mafia festivals, such as the *Festival della Legalità*, or by promoting the culture of legality.

The Mafia represents the Sicilian Mafia. It is composed of many actors who

(i) Request pizzo from Entrepreneurs,
(ii) Provide benefits to paying Entrepreneurs (e.g., protection from predation, and contract and cartel enforcement), and

[5] Given the lack of an English word that means extortion money paid to mafia, we employ the Italian term 'pizzo' that has this meaning.

(iii) Punish non-paying and reporting ones with a specific severity. They are coordinated in their actions—whom they target, how often they request pizzo, how much they request, and how severely they punish—because they are part of the same family. Mafiosi can

(iv) Turn pentiti (a very unlikely event) and help the State capture other Mafiosi, and

(v) Mafiosi who are captured by the State are temporarily removed from the simulation and may provide information about other Mafiosi and the Entrepreneurs who paid pizzo to it in the past allowing the State to approach these Entrepreneurs for evidence. Mafiosi are linked to one another via a scale-free network.

Consumers are multiple actors who do not directly interact with the Mafia. They are connected to other Consumers and Entrepreneurs in a scale-free network; this determines the other actors with which they socially interact. Consumers have the goal to purchase a product and their single decision is to (i) buy a product from Entrepreneurs. The decision regarding which Entrepreneur to buy from is based on a combination of economic considerations (i.e., price of the product) and normative considerations (i.e., relative strength of the norm of buying from Entrepreneurs who do not pay pizzo, dynamically updated over the simulation). They serve as (ii) 'reservoirs' of normative attitudes and behaviours and automatically (iii) spread information that can influence other Consumers and Entrepreneurs.

The Intermediary Organisation is a single actor that embodies a civil society or business organisation. It (i) promotes the culture of legality among Entrepreneurs and Consumers through events such as talks in schools, or the organisation or participation in festivals: for instance the civic organisation *Libera* is the main organiser for the aforementioned *Festival della Legalità*. It (ii) serves as an organisation that Entrepreneurs can join if they are not paying pizzo.

The decisions making of actors in the Palermo Scenario can be broadly divided into two different levels of complexity. The Entrepreneurs and Consumers are endowed with complex decision making abilities and base their choices on a combination of economic and social norm based reasoning, whereas, the State, the Mafia, and the Intermediary Organisation are represented as reactive actors whose decisions are based on fixed probabilities that are initialised at the start of the simulation.

The Entrepreneurs' and Consumers' decisions are taken assuming that the utility of an actor consists of an 'individual' component, which represents the economic part of their reasoning, and a 'normative' component, which represents the social norm based aspect. The individual component ΔIapproximates instrumental decision-making and involves strict cost-benefit calculations that motivate actors to take decisions that maximise their own direct utility. It models actors' motivation to maximize their own utility, independently of what a certain norm dictates. The normative component ΔN models the actor's motivation to comply with a norm. It is a function of *norm salience*; a parameter updated by each actor based on its own behaviour and the information gathered by observing the behaviour of other actors.

Following Conte et al. (2014, p. 99), we use 'norm salience' to refer to a measure that indicates how active and prominent, or inactive and inconspicuous, a norm is within a group in a given context. Formally,

$$Sal^n = \frac{1}{\alpha}\left(\beta + \left(\begin{array}{l} \frac{C-V}{C+V} \times w_c + \frac{O_c - O_v}{O_c + O_v} \times w_o + \frac{\max\left(0,\left(O_v + V\right) - P - S\right)}{O_v + V} \times w_{npv} \\ + \frac{P \times w_p + S \times w_s}{\max(P + S, O_v + V)} + \frac{E_c - E_v}{E_c + E_v} \times w_e \end{array}\right)\right)$$

where, n is the norm being evaluated; α is a normaliser that renders the salience in the range [0,1]; C is the number of times the actor complied with the norm n; V is the number of times the actor violated the norm n; O_c is the number of times the actor observed other actors complying with the norm n; O_v is the number of times the actor observed other actors violating the norm n; P is the number of punishments received, applied or observed due to the violation of norm n; S is the number of sanctions received, applied or observed due to the violation of norm n; E_c is number of messages received from others 'demanding' that the actor complies with the norm n; and E_v is number of messages that the actor received 'demanding' the violation of the norm n.

Each term in the norm salience calculation has a weight value associated with it, and the coefficients α and β have the values 6.27 and 2.97, respectively. These are used to assign different importance to each of the factors in generating the overall norm salience. In Table 4.1, the weight associated to each term is presented, the values of which are based on Cialdini et al.'s (1990) work. It is important to stress that the important aspect of these weights is the proportionality among them and not their specific value.

The specific social norms that Entrepreneurs and Consumers consider are shown in Table 4.2.

$N1_T$ and $N1_N$ are norms that potentially influence the decision of Entrepreneurs to pay extortion money to Mafiosi following a request, and $N2_T$ and $N2_N$ are norms that can play a role in Entrepreneurs' decision to report the request for extortion money by Mafiosi to the State. N3 is a norm that can influence the Consumers' decisions regarding which Entrepreneur to purchase a product from.

Norms $N1_T$ and $N2_T$ are part of the set of norms that are associated with the traditional mentality of the individuals regarding the Mafia in Sicily: for instance, Mafiosi should be paid and not reported to the police (*omertà*). We refer to this set of social

Table 4.1 Social cues and weights for the Norm Salience updating. (Andrighetto et al. 2010)

Cue	Description	Weight
C/V	Own Norm Compliance/Violation	w_c=(+/−) 0.99
O	Observed Norm Compliance	w_o=+0.33
NPV	Non-Punished Violators	w_{npv}=−0.66
P	Observed/Applied/Received Punishment	w_p=+0.33
S	Observed/Applied/Received Sanction	w_s=+0.99
E	Observed/Applied/Received Norm Invocation	w_e=+0.99

Table 4.2 Social norms influencing actors' behaviour in the Palermo Scenario

ID	Norm	Ruled Actor	Content of the Norm
$N1_T$	Pay pizzo	Entrepreneur	Pay money to Mafioso after request
$N1_N$	Do not pay pizzo	Entrepreneur	Do not pay money to Mafioso after request
$N2_N$	Report pizzo	Entrepreneur	Report requests for money
$N2_T$	Do not report pizzo	Entrepreneur	Do not report requests for money
N3	Do not buy from paying pizzo Entrepreneurs	Consumer	Do not buy products from Entrepreneurs known to pay extortion money to Mafiosi

norms as TRADITIONAL. An Entrepreneur is said to hold the TRADITIONAL set of norms if the norm salience value of norms $N1_T$ and $N2_T$ are respectively higher than the norm salience value of norms $N1_N$ and $N2_N$. Conversely, norms $N1_N$ and $N2_N$ represent the NEW set of norms that correspond to a recent emerging anti-mafia sentiment that is based on the understanding of the social and economic harm caused by the Mafia. An Entrepreneur can be said to hold the NEW set of norms if the norm salience value of norms $N1_N$ and $N2_N$ are respectively greater than the norm salience value of norms $N1_T$ and $N2_T$. Differently to these, norm N3 is one factor that is used by Consumers to rank the different Entrepreneurs that may buy a product from.

Although our model is primarily based on the state of affairs that occurred, and is occurring, in Palermo, essentially all of the key ingredients that we identify and implement are present in other mafias and the systems that they are a part of. The set of agents and their relationships implemented here can be used to examine different variants of the same phenomenon.

4.3 Experiment

In this section, we describe a simulation experiment[6] aimed at understanding the dynamics of mafias, and examining the effects of different policies in favouring their destabilisation. More specifically, it enables us to evaluate the effects that the independent and combined actions of the different actors have on mafias. We contend that both legal and social norms are individually important for destabilising and undermining mafias; however, when aligned they are more effective.

The simulation experiment consists of five separate configurations, each of which represents a significant period in the history of the approaches employed by the key actors in Sicily regarding the mafia phenomenon. The set of policies that the actors can use in our experiment are displayed in Table 4.3.

These policies are linked to the simulation model through different input parameters as shown in Table 4.4.

[6] The simulator used to perform this experiment is found at https://github.com/gnardin/gloderss.

Table 4.3 Description of the actors' possible policies

Actor	Policy name	Description
State	Legal/punitive	Represents the use of coercive instruments to fight the Mafia. It is measured in different levels and we classify them as *Weak* or *Strong*. Weak means that the State does not use efficiently or does not have legal or material mechanisms to countering the Mafia. Strong means the State is efficient in using the available resources to fight the Mafia
	Moral Suasion	It is the ability of the State of promoting a culture of legality and to persuade and attract citizens by legitimacy of policies and the values and norms underpinning them
Mafia	Strategy	It corresponds to the strategy used by the Mafia to impose its will. The Mafia has two possible strategies, *Violent* and *Hidden*. The Violent strategy is characterised by the demand of very high amounts of money as extortion, and the infliction of a high and certain punishment on those that do not comply with its extortion request. Conversely, the Hidden strategy is characterised by demanding a low amount of money as extortion from a larger number of Entrepreneurs, and refraining from hardly punishing those that do not pay in order to avoid undercover of its activities by the State
Intermediary Organisation	Social Norm	The spreading of social norms intends to promote the culture of legality among the civil society. It is comparable to the State Moral Suasion, but performed by civil society organisations

Table 4.4 Policies' input parameters

Policy	Input parameter	Description
State Legal/punitive	numPoliceOfficers	Number of police officers
	captureProb	Probability of capturing a Mafioso if the State observes the Mafioso requesting money or punishing Entrepreneurs
	convictionProb	Probability of convicting a Mafioso after it is captured
	percTransferFondo	Percentage of Mafiosi's confiscated resources allocated into a fund supporting the victims of the Mafia
State Moral Suasion	propCitizens	Proportion of the population to receive a message invoking the New set of norms and information about actions of the State countering the Mafia
Mafia Strategy	extortLevel	The amount of the Entrepreneurs' endowment requested as pizzo
	punishSeverity	The amount of punishment effectively inflicted by the Mafiosi on the Entrepreneur that did not pay the extortion request
	punishProb	Probability of punishing a non-paying Entrepreneur
Intermediary Organisation Social Norm	propCitizens	Proportion of the population who receive a message invoking the New set of norms

The simulation experiment is comprised of five configurations that represent identifiable periods in the development and history of the Sicilian Mafia: before 1980, between 1980 and 1992, between 1992 and 1995, between 1995 and 2000, and after 2000. In each of these periods, the actors' policies vary by just one significant feature following the broad historical reality. The specific combinations of the actors' policy features that are used in each of the scenarios are shown in Table 4.5.

Configuration *S1* represents the situation before 1980s, in which the Italian State had few specific legal mechanisms to fight the Mafia (State uses Weak Legal/punitive and No Moral Suasion). The Mafia, conversely, demanded a high amount of money from Entrepreneurs, and Entrepreneurs who did not comply were certainly punished (Violent Mafia). Additionally, most of the population still has a traditional view on the Mafia, in which the payment of systemic extortion is perceived as a "legitimate" retribution for protection services (Gambetta 1993; Varese 2013).

Configuration *S2* represents the 1980s, in which the Italian State instituted several new coercive laws in order to characterise and counter the Mafia. These new

Table 4.5 Configurations' policy features representing distinct historical periods of the Sicilian Mafia in which policies vary from previous configuration by one feature (italic)

Configuration	Period	Actor	Policy value
S1	Pre-1980	State	Weak Legal/punitive
			Inactive Moral Suasion
		Mafia	Violent
		Intermediary Organisation	No Active
S2	1980–1992	State	*Strong Legal/punitive*
			Inactive Moral Suasion
		Mafia	Violent
		Intermediary Organisation	No Active
S3	1992–1995	State	Strong Legal/punitive
			Inactive Moral Suasion
		Mafia	*Hidden*
		Intermediary Organisation	No Active
S4	1995–2000	State	Strong Legal/punitive
			Inactive Moral Suasion
		Mafia	Hidden
		Intermediary Organisation	*Active*
S5	Post-2000	State	Strong Legal/punitive
			Active Moral Suasion
		Mafia	Hidden
		Intermediary Organisation	Active

laws rendered the State institutions (i.e., Police and Judiciary) more effective in countering the Mafia and also in providing support to its victims. These changes rendered the State stronger in directly countering the Mafia (State uses Strong Legal/punitive). However, the State still does not promote a culture of legality by other means (No Moral Suasion).

In 1992 (Configuration *S3*), however, due to the improved effectiveness of the State policy, the Mafia changes its violent and combative strategy (Mafia Violent) into a more moderate strategy in order to operate hidden from the law enforcement (Mafia Hidden). Concretely, this strategy reduces the amount of demanded pizzo, but comprises the request from a larger number of Entrepreneurs, and inflicts a lower punishment in those that do not pay. These changes, especially the State changes, paved the way for the emergence of civil organisations (i.e., Intermediary Organisations) responsible for promoting the culture of legality among the population (i.e., Entrepreneurs and Consumers) beginning in the middle of the 1990s (Configuration *S4*).

After 2000 (Configuration *S5*), the State realized that legal mechanisms were not sufficient to counter the Mafia and it began to act in order to also promote a culture of legality (State uses Moral Suasion). This is reflected on supporting initiatives to promote these values in schools and among the general public (e.g., *Festival della Legalità*).

The input parameters defining these configurations are shown in Table 4.6.

The input parameters' values associated to the actors' policy features have been extracted from empirical work conducted in Sicily by the GLODERS[7]

Table 4.6 Input parameter values

Policy	Value	Input parameter value
State Legal/punitive	Weak	numPoliceOfficers=5
		captureProb=0.2
		convictionProb=0.1
		percTransferFondo=0.0
	Strong	numPoliceOfficers=20
		captureProb=0.8
		convictionProb=0.6
		percTransferFondo=0.5
State Moral Suasion	Inactive	propCitizens=0.0
	Active	propCitizens=0.05
Mafia Strategy	Violent	extortLevel=0.1
		punishSeverity=0.75
		punishProb=0.9
	Hidden	extortLevel=0.03
		punishSeverity=0.5
		punishProb=0.5
Intermediary Organisation Social Norm	Inactive	propCitizens=0.0
	Active	propCitizens=0.1

[7] http://www.gloders.eu.

(Global Dynamics of Extortion Racket Systems) partner affiliated with the University of Palermo (Militello et al. 2014; La Spina et al. 2014). These data were collected through interviews of extorted entrepreneurs, judicial documents and confiscated Mafia documents analyses (e.g., the *Libro Mastro*).

4.4 Results

For each configuration, the simulation model was run with 200 Consumers, 100 Entrepreneurs and 20 Mafiosi. The number of Police Officers varies depending to the policy adopted by the State (see numPoliceOfficers parameter value in Table 4.6). The analyses of the configurations (see Table 4.5) are based on a set of output metrics described in Table 4.7.

Figure 4.1 illustrates the results obtained by simulating the distinct historical periods.

Figure 4.1a shows that in the transition from the configuration S1, in which the State has a Weak Legal/punitive, to configuration S2, in which the State has a Strong Legal/punitive policy, the number of extortion requests reduces dramatically. Such result is supported by empirical evidence that indicates the high influence of coercive mechanisms used by the State in countering this type of criminal organisations. Inspecting the proportion of paid pizzo requests (Fig. 4.1b), however, we observe that even though the number of pizzo requests has decreased, the State is not effective in preventing Entrepreneurs from paying pizzo as there is almost no difference in the proportion of paid pizzo requests between configurations S1 and S2. The same is noticed in the proportion of reports (Fig. 4.1c), which in both configurations is very low.

Figure 4.2 (see values in Table 4.8) shows the Entrepreneurs' norms salience mean value. In scenarios S1 and S2, we observe the same norms' salience changing pattern, in which the salience of the TRADITIONAL set of norms increases. It means that the population becomes more inclined to comply with and not report pizzo requests. This suggests that even though the State is effective in capturing and convicting Mafiosi (see Fig. 4.1d), which results in a drastic reduction in the number of pizzo requests (see Fig. 4.1a), it is not successful in changing the population's behaviour regarding the payment and reporting of pizzo requests.

Table 4.7 Output metrics

Metric	Description
Number of extortions	Total number of pizzo requests
Proportion of paid extortion	Proportion of pizzo requests paid by the Entrepreneurs
Proportion of reports	Proportion of reports to the State
Proportion of imprisonments	Proportion of investigations leading to imprisonments

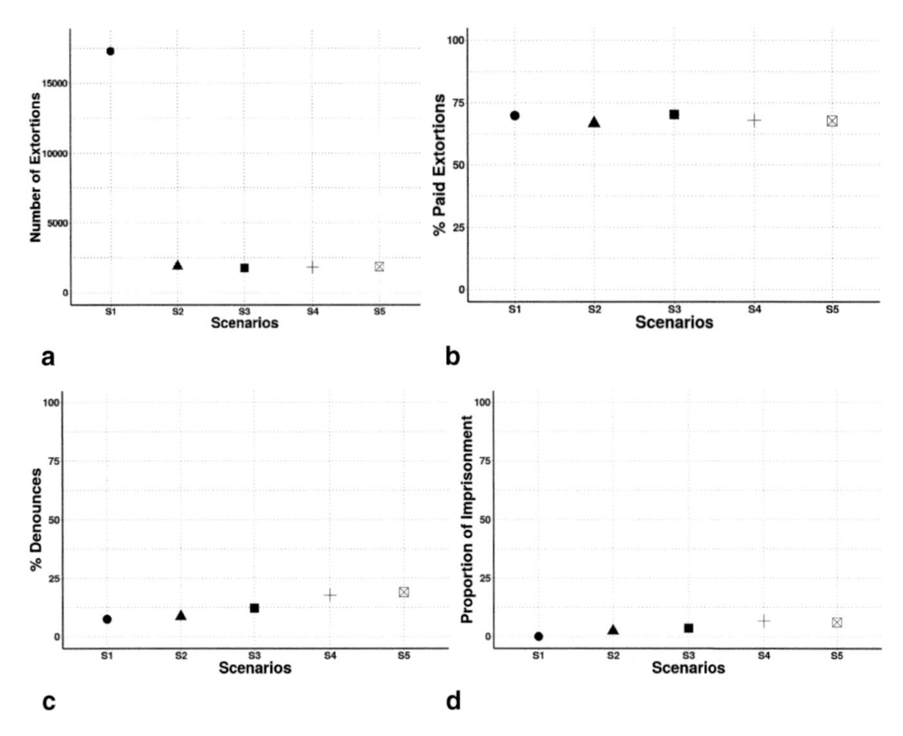

Fig. 4.1 Simulation results of the historical Palermo Scenario configurations. **a** Total number of pizzo requests, **b** Proportion of paid pizzo requests, **c** Proportion of reporting, **d** Proportion of imprisonments

In configuration S3, the Mafia strategy changes to a strategy of submersion characterised by lower requests and softer punishment. As a result of this new strategy, the Mafia is successful in recovering the proportion of paid pizzo requests lost in S2. This new strategy of inflicting less punishment on non-payers, however, has the unpredicted effect of rendering the act of reporting more attractive for the Entrepreneurs. The success of the new Mafia strategy may be, in part, imputed to the TRADITIONAL set of norms still highly salient in the majority of the population and thus to the inadequacy of the State in favouring a change towards the NEW set of norms. This is observable in Fig. 4.2, in which the salience of the TRADITIONAL and NEW set of norms remains relatively unchanged in configuration S3.

This assumption is tested by the inclusion of a new actor (i.e., the Intermediary Organisation) in configuration S4, whose main activity is to promote the NEW set of norms. As a result, in Fig. 4.1) we observe a reasonable reduction in the proportion of paid pizzo requests and also a change towards an increase of the salience of the NEW set of norms.

More interesting yet is the high number of actors (about 11.0 %) that shifted from the TRADITIONAL to the NEW set of norms (see Table 4.7). In addition, another 19.6 % of actors shifted the norm regarding the payment of pizzo requests (from a higher salience of the norm ‘Pay pizzo’ to a higher salience of the norm ‘Do not pay pizzo’), but it did not change the norm regarding reporting pizzo requests. Thus the

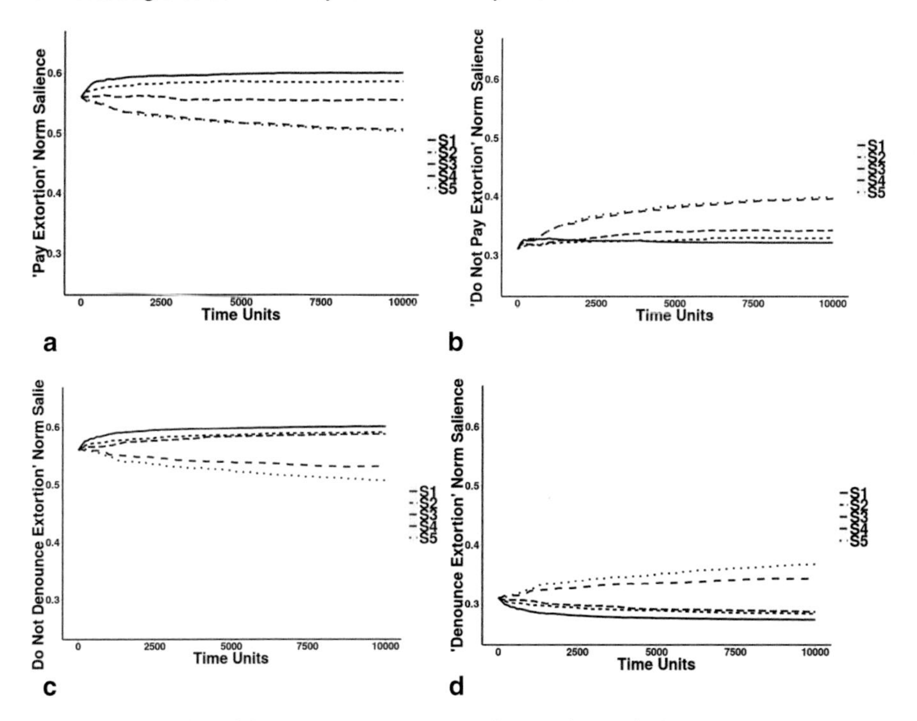

Fig. 4.2 Mean value of the Entrepreneurs' norms salience. The y-axis show the mean strength of the norms in the whole population and the x-axis represent the elapsed simulation time measured in time units. **a** 'Pay pizzo' norm ($N1_T$), **b** 'Do not pay pizzo' norm ($N1_N$), **c** 'Do not report pizzo' norm ($N2_N$), **d** 'Report pizzo' norm ($N2_T$)

Table 4.8 The norm salience's mean value of the norms at the end of the simulation

Norm	Configurations				
	S1	S2	S3	S4	S5
$N1_T$	0.599 ± 0.06	0.585 ± 0.05	0.554 ± 0.03	0.506 ± 0.06	0.503 ± 0.07
$N1_N$	0.321 ± 0.01	0.329 ± 0.03	0.342 ± 0.03	0.396 ± 0.10	0.398 ± 0.10
$N2_N$	0.272 ± 0.03	0.282 ± 0.03	0.285 ± 0.02	0.343 ± 0.05	0.367 ± 0.06
$N2_T$	0.601 ± 0.03	0.591 ± 0.03	0.588 ± 0.02	0.530 ± 0.05	0.507 ± 0.06

promotion of the culture of legality performed by the Intermediary Organisation proves effective in changing the Entrepreneurs normative mind-set, which is also reflected in the reduction of paid pizzo requests and the increase in reporting.

Finally, in configuration S5, the State begins an activity that complements the action of the Intermediary Organisation in promoting a culture of legality, by encouraging the adoption of the NEW set of norms and by giving more visibility to the actions taken and the results obtained in countering the Mafia (Moral Suasion). Looking at the graphics in Figs. 4.1 and 4.2, we note no significant change with respect to values of configuration S4.

Nonetheless, analysing the transitions shown in Table 4.9, we find that a greater number of actors adopt the NEW set of norms, about 20.7 %. Even though a clear

Table 4.9 Proportion of actors with certain norms in the end of the simulation in relation to the beginning of the simulation

Configuration	Traditional (%)	Only 'Do not pay pizzo' (%)	Only 'Report pizzo' (%)	New (%)
S1	100.0	0.0	0.0	0.0
S2	100.0	0.0	0.0	0.0
S3	99.3	0.7	0.0	0.0
S4	68.7	19.6	0.7	11.0
S5	67.0	11.3	1.0	20.7

behavioural change is still not observable, we expect that a higher number of actors with the NEW set of norms will result in a more resilient system.

4.5 Conclusions

This paper proposes an agent-based simulation model aimed at understanding how social processes may sustain legal processes, enacted by the State, in counteracting the Mafia and its extortion activities. The model is populated by agents with normative capabilities that allow them to autonomously detect and reason on social norms. These features allow the inspection of their mind and even the manipulation of unobservable variables.

Those inspections enabled us to identify that even though legal norms are effective in fighting the Mafia and changing the Entrepreneurs' behaviour regarding the payment of pizzo requests; it is not effective in changing their normative beliefs about pizzo. This results in a fragile change that looking only at the behaviour of the Entrepreneurs is undetectable. However, by comparing the proportion of pizzo requests payment in configurations S2 and S3, we could observe that a simple change of the Mafia strategy neutralised any gains obtained by the State actions. Whenever the Entrepreneurs social norms are changed to or towards the NEW set of norms, we observe that they become more resistant to the Mafia requests. Hence, we can conclude that in order to legal norms to become resilient and effective, they must be supported by social norms; otherwise the achieved change is weak.

These strengths enable us to gain a deeper understanding of mafias and support us in answering the research questions posed in Sect. 1:

1. *What effects do different policies of the state have on destabilising mafias?*

The State legal/punitive policy is effective in countering the Mafia and reducing its extortion activities, yet it alone is not able to change the population's mind-set from the TRADITIONAL to the NEW set of norms (i.e., culture of legality). This results in a fragile situation in which the population will switch back to the previous behaviour of complying with the Mafia requests if the State reduces or stops its effort in acting coercively against the Mafia.

The State Moral Suasion policy, apparently, does not show large benefits in changing the population's behaviours; however, when it supports the activity of the Intermediary Organisations in favouring the spread of a culture of legality, the State Moral Suasion policy favours the norm shift from the TRADITIONAL to the NEW set of norms.

2. *How do independent and combined actions of different actors (such as the state, non-governmental organisations, and civil society) affect mafias?*

The policies of the State and the Intermediary Organisation are complimentary. The coercive policies used by the State against the Mafia helps to reduce the Mafia's activity. However, the change that it generates is fragile as any reduction in the State's coercive activity may allow the Mafia to re-emerge and re-impose its requests and power on the population without any resistance from it. In contrast, the Intermediary Organisation's activities to promote a culture of legality promotes a shift in the population's normative mind-set, which may render a return of extortion activities a more difficult task as the population will be more prone to report and resist pizzo requests.

3. *Which conditions favour the spread of the culture of legality that undermine mafias?*

Our results show that promoting a culture of legality through the spreading of normative information improves the shift from the TRADITIONAL to the NEW set of social norms.

Interesting future work includes (i) validating the simulation model to different Italian provinces and countries; (ii) evaluating the impact of other policy combinations in countering the mafias; and (iii) testing the effects of external sudden changes (such as an economic crisis, or a sudden change in the policy of the State) on the dynamics of the simulation scenarios.

Acknowledgements This work was partially supported by the FP7-ICT Science of Global Systems programme of the European Commission through project GLODERS (http://www.gloders.eu, "Global Dynamics of Extortion Racket Systems") under grant agreement no.: 315874. This work reflects solely the views of its authors. The European Commission is not liable for any use that may be made of the information in the work.

References

Andrighetto, G., Villatoro, D., & Conte, R. (2010). Norm internalization in artificial societies. *AI Communications, 23*(4), 325–339.

Barone, G., & Narciso, G. (2013). The effect of mafia on public transfers. Working paper series, The Rimini Centre for Economic Analysis.

Bicchieri, C. (2006). *The grammar of society: The nature and dynamics of social norms*. New York: Cambridge University Press.

Cialdini, R. B., Reno, R. R., & Kallgren, C. A. (1990). A focus theory of normative conduct: Recycling the concept of norms to reduce littering in public places. *Journal of Personality and Social Psychology, 58*(6), 1015–1026.

Conte, R., Andrighetto, G., & Campennì, M. (Eds.). (2014). *Minding norms: Mechanisms and dynamics of social order in agent societies. Oxford series on cognitive models and architectures*. Oxford: Oxford University Press.

Daniele, V. (2009). Organized crime and regional development. A review of the Italian case. *Trends in Organized Crime, 12*(3–4), 211–234.

Elster, J. (2007). *Explaining social behavior: More nuts and bolts for the social sciences* (2rev ed.). Cambridge: Cambridge University Press.

Elster, J. (2009). Norms. In P. Hedström & P. Bearman (Eds.), *The Oxford handbook of analytical sociology* (pp. 195–217). Oxford: Oxford Univiversity Press.

Gambetta, D. (1993). *The Sicilian mafia: The business of private protection*. Cambridge: Harvard University.

Hill P. B. E. (2006). *The Japanese mafia: Yakuza, law, and the state*. Oxford: Oxford University Press.

La Spina, A., Frazzica, G., Punzo, V., & Scaglione, A. (2014). How mafia works. An analysis of the extortion racket system. In Proceedings of ECPR General Conference.

Lavezzi, A. M. (2008). Economic structure and vulnerability to organised crime: Evidence from Sicily. *Global Crime, 3*(9), 1998–1220.

Militello, V., La Spina, A., Frazzica, G., Punzo, V., & Scaglione, A. (2014). Quali-quantitative summary of data on extortion rackets in Sicily. Deliverable 1.1, GLODERS Project.

Morgan, W. P. (1960). *Triad societies in Hong Kong*. Hong Kong: The Government Printer.

Nardin, L. G., Andrighetto, G., Conte, R., Székely, Á., Anzola, D., Elsenbroich, C., Lotzmann, U., Neumann, M., Punzo, V., & Troitzsch, K. G. (under review). Simulating the dynamics of extortion racket systems: A Sicilian mafia case study. *Journal of Autonomous Agents and Multi-Agent Systems*.

Pinotti, P. (2012). *The economic costs of organised crime: Evidence from Southern Italy*. Technical report, Banca d'Italia, Roma, Italy.

Savona, E. U. (2012). Italian mafias' asymmetries. In D. Siegel & H. van de Bunt (Eds.), *Traditional organized crime in the Modern World, volume 11 of studies of organized crime* (pp. 3–25). New York: Springer.

Schelling, T. C. (1967). The strategy of inflicting costs. In R. N. McKean (Ed.), *Issues in defense economics* (pp. 105–128). New York: National Bureau of Economic Research, NBER Books.

Schelling, T. C. (1971). Dynamic models of segregation. *Journal of Mathematical Sociology, 1*, 143–186.

Varese, F. (1996). What is the Russian mafia? *Low Intensity Conflict and Law Enforcement, 5*, 129–138.

Varese, F. (2001). *The Russian mafia: Private protection in a new market economy*. Oxford: Oxford University Press.

Varese, F. (2013). *Mafias on the move: How organized crime conquers new territories*. Princeton: Princeton University.

Zhang, S. (2012). China tongs in America: Continuity and opportunities. In D. Siegel & H. van de Bunt (Eds.), Traditional organized crime in the modern world, volume 11 of studies of organized crime (pp. 109–128). New York: Springer.

Chapter 5
Experimental Economies and Tax Evasion: The Order Beyond the Market

Juliana Bernhofer

In this world nothing can be said to be certain, except death and taxes.
Benjamin Franklin

5.1 Introduction

Tax non-compliance and fiscal fraud are giving policymakers all over the world quite a headache, leading them to invest significant efforts and resources in the attempt to tackle the issue of public revenues lost due to tax evasion.

To achieve meaningful results through targeted policy interventions, it is central to understand how people perceive taxes, contributions, and sanctions, along with how they take the decision to comply. Various attempts have been made, starting from the seminal contributions of Allingham and Sandmo (1972) and Yitzhaki (1974), who follow the paradigms of expected utility theory. These two classical theoretical models are presented in Sect. 5.2, as they serve as a starting point for the majority of further developments in the tax-compliance literature. Despite their elegance and insightfulness, the classical models were soon questioned as it was shown that their predictions did not match existing empirical evidence (see, for example, Graetz and Wilde 1985). In fact, people were found to evade much less than what would have been expected from a rational utility maximizer with a reasonable level of risk aversion. Sections 5.3 and 5.4 deal with issues related to the assumption of full rationality and the potential problems arising with the use of expected utility theory.

J. Bernhofer (✉)
Department of Economics, Ca' Foscari University of Venice,
Cannaregio 873, 30121 Venice, Italy
e-mail: juliana.bernhofer@unive.it

F. Cecconi (ed.), *New Frontiers in the Study of Social Phenomena*,
DOI 10.1007/978-3-319-23938-5_5

The process of cross-checking real data with model prescriptions and the acknowledgment of the limitations of analytic solutions led to modifying the basic research question from, "Why and by how much do people evade taxes?" to "Why do people pay taxes at all?" The answer should ultimately enable policy makers to find a solution to their old dilemma of, "How can compliance be increased even more?"

However, determining the extent of tax evasion is and has always been an understandably challenging task. There are four main methods, according to Andreoni et al. (1998): audit data, survey data, tax amnesty data, and laboratory experiments.

A frequently cited source associated with the first category is the Tax Compliance Measurement Program (TCMP) carried out by the U.S. Internal Revenue Service (IRS) from 1965 to 1988. The TCMP was a program of intensive audits designed to measure the level of noncompliance among the population. The results of the 1988 TCMP show that 40 % of the households evaded some income tax, whereas 53 % were fully compliant. The remaining 7 % instead paid more than what they actually owed.

Yet available data on tax compliance is limited and often unreliable. Audit data does not give sufficient insights, as it is nearly impossible to detect all hidden income. Survey data, on the other hand, is self-reported, which casts reasonable doubt on the truthfulness of the information provided about one's own illicit behavior. To overcome these and other limitations, experimental economics comes into play in the attempt to provide at least a partial fix. By carefully constructing a laboratory environment as described by an economic model (e.g., tax rate, audit frequency, fine rate), the experimenter may observe whether participants behave according to the analytical predictions. Furthermore, marginal effects of single-parameter changes can be isolated *ceteris paribus*, thanks to the controlled setting of the lab. In Sect. 5.5, we will present some results obtained from laboratory experiments on tax compliance.

The body of literature on tax evasion has grown to massive proportions during the last three decades and has become cumbersome to overlook. However, two of the main common findings are that individuals are not homogeneous in preferences and often do not act according to market-based mechanisms. Yet the plurality of stylized determinants of tax compliance and the discovery of the importance of interaction effects and dynamic approaches have given rise to new ways of modeling. Some examples of computer-simulated agent-based models applied to the tax evasion problem are illustrated in Sect. 5.6. The potential benefit of calibrating such models with the results stemming from real "agents" tested in controlled laboratory settings still has to be further explored. This new area of interaction between both Experimental Economics and computer-simulated realities—or an Experimental Economy—could well be one innovative tool to bridge the gap between theoretical modeling and reality on one side, while decreasing the distance between academics and policy makers on the other, thanks to the creation of manipulable interfaces that are also intuitive for non-technical users.

5.2 Classic Modeling of Tax Evasion

A classic approach to the problem of tax evasion is offered by the theoretical model contained in the seminal article of Allingham and Sandmo (1972) (henceforth AS). The model is an adaptation of Gary Becker's work on the Economics of Crime and Punishment (Becker 1974) to the case of tax compliance. The decision on how much to declare of a certain exogenously given income is presented as a gamble with two possible outcomes: being caught and not being caught. The tax authority audits the taxpayer and discovers the understatement with a certain probability p; hence the decision-maker maximizes his expected utility (EU) with respect to the declared amount x according to the following convex combination:

$$EU = (1-p)U\left(v+t\left(y-x\right)\right)+pU\left(v-\theta\left(y-x\right)\right) \tag{5.1}$$

His overall expected utility is represented as a weighted average of the utilities assigned to the two possible outcomes. The optimum amount of declared income x depends on the proportional tax rate t, the fine rate θ, and the probability of being subject to a random audit p. The utility function $U(.)$ is marginally positive with $U'(.) > 0$ and strictly decreasing with the second order derivative $U''(.) < 0$, which means that the taxpayer is a risk-averse one who prefers a certain outcome to a gamble with the same (or even a higher) payoff in expected terms.

The correct net disposable income in case of full income disclosure is described by $v = y(1-t)$. Thus one can interpret the first part of the expected utility formulation as the situation in which no audit is performed and the utility is given by the argument $(v+t(y-x))$, the correct net income v augmented by what we will call the *cheater's premium*, $t\left(y-x\right)$, namely, the part of tax liability he saved by not paying taxes on the undeclared part. The second part of the weighting function describes the situation in which an audit takes place: the undeclared income $\left(y-x\right)$ is detected by the authority and the fully taxed correct net income v is reduced by the *cheater's penalty* $\theta\left(y-x\right)$, i.e., the fine on the hidden income.

Commonly, decreasing absolute risk aversion is assumed, which describes the fact that—in absolute terms—the amount of risky investments increases with higher disposable income.

The first-order condition (FOC) for the optimal amount of income disclosure[1] of Eq. (5.1) is $\frac{\partial EU}{\partial x} = 0$ and becomes $\frac{U'\left(y_A\right)}{U'(y_{NA})} = \frac{\left(1-p\right)t}{p\theta}$. We find that in the AS specification an increase in the tax rate t has ambiguous effects. On one hand, the correct net disposable income v decreases, which—under the assumption of decreasing absolute risk aversion[2]—should induce the individual to cheat less, leading

[1] The second-order condition is satisfied by the utility function's being concave.

[2] The concepts of absolute and relative risk aversion have been developed by Arrow (1965) and Pratt (1964) independently. The first measure describes the amount of wealth placed in risky activities in absolute terms and the second expresses this amount in relative percentage terms.

to an income effect. On the other hand, with a higher tax rate, the attractiveness of the *cheater's premium* $t(y-x)$ increases, whereas the *cheater's penalty* $\theta(y-x)$ remains unaffected, which eventually makes tax evasion more profitable (substitution effect). The magnitude and sign of the final response depend on the shape of the utility function, in particular on how fast absolute risk aversion declines and thus its third-order derivative (Andreoni et al. 1998).

With the aim of reflecting the legal framework effective in a number of countries, such as the United States and Israel, in 1974, Shlomo Yitzhaki (1974) introduced a slight modification to the specification of the AS model by making the penalty depend on the evaded taxes instead of the undeclared amount of income. The original expected utility expression becomes

$$EU = (1-p)U\left(v + t(y-x)\right) + pU(v - \theta t(y-x)) \tag{5.2}$$

and we find the following first-order conditions for optimality: $\frac{U'(y_A)}{U'(y_{NA})} = \frac{(1-p)}{p\theta}$. The taxpayer will evade as long as the expected payoff per monetary unit of evasion $1-p-p\theta$ is greater than zero.

By directly comparing the F.O.C. of Eq. 5.2 with the original AS first-order condition, we observe that the proportional multiplicative effect of taxes in the numerator, which made evasion more attractive, disappears and what remains is only the income effect (Slemrod and Yitzhaki 2002). Overall, cheating will be reduced when the tax rate increases. Moreover, both the probability of an audit and the magnitude of the fine have a negative impact on evasion.

The Allingham–Sandmo–Yitzhaki (henceforth ASY) model stands out for its elegance and straightforwardness, and thus became the standard tool, or at least the starting point, in the analysis of the compliance decision. Yet an ever-growing body of empirical and experimental evidence has developed pointing out that the predictions of the ASY model do not fully hold up in reality. In the following sections we will highlight some of the criticisms to this classic utilitarian approach.

5.3 Limits of Rational-Choice Theory

Presenting the decision of tax compliance as an individual-choice problem in order to resolve what could be well defined as an aggregate social problem has proven to lack significant elements that influence the process of human decision-making. The rational choice approach has been widely challenged, as described also by the American economist James M. Buchanan:

> The economist rarely examines the presuppositions of the models with which he works. The economist simply commences with individuals as evaluating, choosing, and acting units. Regardless of the possible complexity of the processes or institutional structures from which outcomes emerge, the economist focuses on individual choices. [...] Individuals [...] are presumed able to choose in accordance with their own preferences, whatever these may be, and the economist does not feel himself obliged to inquire deeply into the content of these preferences (the arguments in individuals' utility functions). (Buchanan 1987, p. 244)

Standard neoclassical models of economic theory are built on the assumption of individuals' exhibiting rational behavior. Rationality in the economic context is interpreted as the individual's capability of evaluating all possible outcomes in order to take the decision that yields the greatest benefit in terms of utility. Moreover, agents are also assumed to be aware of their own preferences and to be able to maximize their utility function given certain parameters. They do so in a purely self-interested way.

Accordingly, in the classic AS and ASY model of tax compliance the decision of how much to declare from one's income to the tax authority is presented as a relatively simple portfolio choice. Taxpayers must decide how much of their income they wish to allocate to the risky asset (tax evasion) and to the safe asset (tax compliance). Decision-makers are assumed to have full information about the audit probability, the fine, and the tax rate they are supposed to pay and make so-defined "rational choices under uncertainty." The latter is engineered by assigning probabilities to possible outcomes.

Portfolio theory and its basic assumptions of perfect rationality have been widely challenged by the science of behavioral economics, which deals with the social and cognitive aspects in the human decision-making process. By introducing elements of psychology into economic modeling, new points of view have been presented which are not necessarily in contrast with the neoclassical models.

Extensive experimental research has shown that individuals are only boundedly rational (see, for example, Conlisk 1996 for a survey on bounded rationality). The concept was already introduced in the 1950s by the work of Herbert A. Simon and has since been subsequently defined and modeled by numerous authors (Simon 1982; Selten 2001; Simon 1972; Kahneman 2003). Examples of what could be those bounds to full rationality are information asymmetry and cognitive limitations.

How does bounded rationality apply to the tax evasion problem? First, most of the time citizens do not have complete information about the true audit probabilities and form subjective probability beliefs about the frequency of verifications by the tax authority. Also, tax code complexity and bureaucracy can lead to uncertainty not only about the fine parameter θ, but also about the correct tax rate itself, as pointed out by Andreoni et al. (1998, p. 852). Tax complexity leads to the need for tax practitioners and represents a potential source of inequity among the population, in particular with respect to education and socioeconomic status. These regressive effects stemming from inferior capabilities of interpreting the tax code and, as a consequence, finding ways to minimize the tax liability are also mentioned by Vogel (1974).

Second, optimizing a utility function based on some probabilities of possible outcomes might not be a straightforward process for everyone. It is demonstrated by numerous studies, as described in Reyna and Brainerd (2008), that there is a general lack of mathematical proficiency and subsequent difficulty in judging probabilities and risks among the population. Not only is it numeracy, but also tax literacy that plays a role in determining the correct level of the expected-utility parameters to be taken into consideration.

Tax literacy is tightly linked to the aforementioned tax code complexity. The first is the capability of applying the correct tax rate given a certain legal framework, whereas the latter describes the structure and accessibility of that legal framework. The more complicated the design of the set of rules, the more important the capability to interpret them correctly, in order words, *ceteris paribus*, the marginal return to tax literacy increases.

Finally, it is also assumed that taxpayers take decisions individually and in a self-interested manner, but can we assume this mechanism to be compatible with the very purpose of taxes? Taxes are collected in order to finance public expenditures that again are designed to serve the community of taxpayers. This implies that tax evasion is a form of free-riding. Not paying one's taxes has the effect that the other members of the community have to pay more in order to fund the public collective project. In that sense it might be necessary to consider social interactions and norms in order to capture the role of peer effects, positive and negative reciprocity, and intrinsic motivation in the decision-making process. We will consider these elements more in depth in Sect. 5.5.

5.4 The Expected-Utility Approach Under Scrutiny

In the ASY model the decision to pay taxes is presented as a lottery with two possible outcomes: audited and not audited. Taxpayers are then expected to decide how much to declare based on the probability of being audited and possibly fined. Following this logic and given the population-specific level of risk aversion, it should hence be possible for the lawmaker to provide society with a set of rules defining audit rates, tax rates, and fines which leads to collecting the maximum tax levy in a self-regulatory manner.

Individual heterogeneity is represented with regard to the attitude toward risk and captured by the functional form of the utility function, in particular its curvature. A risk-averse individual is characterized by a concave utility function with a decreasing return to wealth in marginal terms.

The Arrow–Pratt measure of risk aversion in absolute terms describes the relationship between the second-order and the first-order derivative of the utility function, whereas the measure in relative terms describes the level of risk aversion with varying levels of wealth.

The level of risk aversion in the context of tax compliance has been studied by numerous authors. Alm et al. (1992a) showed that estimated Arrow–Pratt levels of relative risk aversion for the United States are incompatible with the empirical evidence of tax compliance. The real levels are between 1 and 2, but only a level of 30 would support observed tax compliance rates. Frey and Feld (2002) find that the observed compliance in Switzerland of 76.52 % would require a value for the parameter of relative risk aversion of 30.75, as opposed to the observed parameter values ranging from 1 to 2.

The empirical calibration values for the model were presented in Alm et al. (1992a), Andreoni et al. (1998), and Bernasconi (1998), with real-world average audit rates ranging between 1 and 3 %, and the penalty rate, which we called θ in our specification, ranging from 0.5 to 2.0. The return to tax evasion in expected terms can be obtained using $1-p-p\theta$ and results in 91–98.5%; hence all taxpayers should hide some of their income, which stands in contrast with the evidence showing that only 30% of taxpayers actually evade taxes Dhami and Al-Nowaihi (2007). Again, only unreasonably high levels of risk aversion could explain the levels of tax compliance found in reality.

Criticisms to the AS and the ASY models of tax evasion link back to the very same discussions around EUT itself. Decision-makers are defined to be (rational) expected utility maximizers if they meet the four basic criteria of the Von Neumann–Morgenstern specification: (i) Completeness—preferences of individuals are well-defined; hence they are able to choose between two alternatives. (ii) Transitivity—the choices are coherent; i.e., if outcome A is preferred to outcome B and B is preferred to C, then it must be that outcome A is preferred to outcome C. (iii) Independence—if gamble A is preferred to gamble B and another gamble C is added to both of them, then preferences do not change: the new gamble (A + C) must still be preferred to the new gamble (B + C). (iv) Continuity—given the preference ranking $A \succ \mathrm{B} \succ \mathrm{C}$ then there must exist some value of p in a convex combination of A and C which makes the decision-maker indifferent to option B, such that $pA+(1-\mathrm{p})\mathrm{C} \approx \mathrm{B}$.

A challenge to the Von Neumann–Morgenstern utility specification is offered by Kahnemann and Tversky and their work on Prospect Theory (Kahneman and Tversky 1979) and Cumulative Prospect Theory (Tversky and Kahneman 1992), which shows inconsistency in preferences describing the nonlinear subjective reaction to probabilities. Starting from a reference income and moving into the gain domain, preferences are concave, whereas in the loss domain preferences are convex. The aim of determining a reference point from which to depart in defining the gain and the loss domain is to eliminate possible framing effects. Going back to our taxation framework, in Dhami and al-Nowaihi (2007) this reference point is defined as the legal after-tax income.

Moreover, concavity of gains and convexity of losses indicates the presence of a loss aversion, where losses are perceived as worse than gains in relative terms. Cumulative Prospect Theory (CPT) uses rank-dependent expected utility theory in order to define the probability weighting function. In that way decision-makers will tend to overweight low probabilities and underweight high probabilities. In taxation terms, such a mechanism implies that a realistic audit rate of, say, 0.01 is subjectively interpreted as higher. Hence compliance for low audit rates increases with respect to the predictions of standard expected utility theory and is more in line with real world data.

This hypothesis was tested in the laboratory by Alm et al. (1992a). In their experiment they set a cut-off level for the audit probability of 5% below which a risk neutral expected utility maximizer should report zero income. Yet at a level of 2% they still find significant compliance rates of around 50%. Such a result could fit expected utility theory only by assuming extreme values of risk aversion. Still, the results are consistent with the predictions of Cumulative Prospect Theory which allows for subjects to subjectively perceive a higher audit rate than the given one.

They also find that the reactions to increases in audit probabilities are non-linear, with compliance rising less than the audit rate in relative terms.

An additional variation of their experiment consisted in a treatment with no possibility of being detected. Nonetheless, the average compliance rate was 20 % which makes the comparison with the neutral lottery set-up somewhat questionable and files Cumulative Prospect Theory as only a piece of the compliance puzzle.

5.5 Institutions, Social Norms and Psychological Factors—New Evidence from the Lab

Indeed, it has become quite clear that there are several further "ingredients" to be considered in order to get a better picture of the various aspects involved in the tax compliance process, aside from the fine rate, the available income, the audit frequency and the tax rate. The *homo economicus*, the economic man, who is assumed to act rationally and in a self-interested manner, would be better off not paying taxes at all if we consider real-world audit rates, even under the assumption of extreme levels of risk aversion.

In this section, we will present some results of tax compliance experiments testing for the impact of both the classical parameters and behaviorally driven elements which have been gradually introduced in the attempt to accommodate observed tax compliance data.

A typical tax compliance experiment is computer-based and consists of one or more rounds during which subjects are asked to take a decision on how much of their previously assigned income they want to declare to the tax authority, given the audit frequency, the fine in cases of detection and the tax rate. Thereafter their report may be randomly drawn for an audit and if they declared less than their gross income, the penalty is applied.

It is also to be mentioned, however, that experiments in economics are often subject to criticism with regards to their external validity. Guala and Mittone (2005) dedicate a section to the issues related to the tax compliance environment in the laboratory, naming as examples problems of scale, the game-like behavior of subjects, the absence of social incentives (the "real" social environment is not part of the experiment) and the absence of social actors. The authors admit to the difficulty of generalizing laboratory results to the real world due to their inherent context-specificity. Nevertheless, tax experiments may offer valuable cause-effect explanations and might often even be the only chance to get additional data on the behavioral dynamics behind the tax compliance decision (Alm et al. 1992b), given the difficulty of gathering truthful and reliable data on tax compliance.

Another criticism common to laboratory experiments in general is the use of student subjects. This is addressed by Alm et al. (2010) and tested in a tax compliance experiment conducted with both, students and staff. Some variation was introduced between groups with regards to the level of certainty about the tax liability and the existence of social programs aimed at positively inducing taxpayers (namely Income Tax Credit and Unemployment Benefits). They find that average compliance rates within subjects (staff and students) indeed differ, but that changes within group treatments are alike. Similarly to Guala and Mittone (2005), their findings suggest that laboratory experiments are able to offer insights regarding marginal effects of parameter changes.

5.5.1 Testing the Classic Microeconomic Predictions

Experiments that test traditional microeconomic models such as those we have seen (AS and ASY), focus on manipulations of the enforcement regime, the tax rate and the fine rate. In an experiment conducted on law students, Friedland (1982) found for example that responsiveness to information about threat probability (audit) is higher than to information about threat magnitudes (fines).

Moreover, a number of experiments have been conducted to assess reactions to variations in the tax burdens. The ASY model predicts the compliance rate to be increasing in the tax rate, however, this cannot be confirmed by a number of findings coming from the lab, which is commonly called the Yitzhaki puzzle. Alm et al. (1992b) find that tax evasion increases with the tax burden which is in-line with the empirical findings of Clotfelter (1983), even with the tax rate elasticity being similar[3]. Also the experimental subjects of Bernasconi et al. (2014) tend to increase their compliance when the tax burden decreases and vice versa. In addition they find that the reaction to tax cuts and tax rises is asymmetric with faster reaction to the first than to the latter.

Cultural factors and social norms might also play a role as shown by Alm et al. (1995), whose results will be presented more in detail in the next section. They find that Spanish test subjects behave according to the ASY model, increasing their compliance with higher tax burdens at a positive rate of 0.94 whereas the U.S. subject pool confirms once again previous findings as in Clotfelter (1983) and Alm et al. (1992b) with a negative elasticity of around −0.5.

An attempt to modify the ASY model in order to match the experimental and empirical evidence has been made by Dhami and Al-Nowaihi (2007) who show that under prospect theory, hence depending on the reference point, tax evasion is increasing in the tax rate[4].

Overall, similarly to empirical results, also in the laboratory higher-than-rational levels of compliance are usually observed, which prompted researchers to investigate the role of determinants other than the classical triad of parameters.

---

[3] around −0.5.

[4] Strictly increasing for interior solutions and non-decreasing in case of boundary solutions when $D^* = 0$ or $D^* = W$.

5.5.2 *Tax Compliance as a Social Norm*

The juxtaposition of the profit-maximizing *homo economicus* in the Smithian sense with Dahrendorf's more other-regarding *homo sociologicus* became a necessary adjustment in the attempt to disentangle the fundamental drivers of the tax compliance decision.

The latter characterization describes an economic agent who acts according to social norms and exhibits feelings such as guilt and anxiety. Roughly said, social norms are behavioral rules shared by other people who tend to judge them in a similar way. Elster (1989) cites some examples of social norms, such as norms of reciprocity, work norms and norms of cooperation. Special forms of norms of cooperation are given by norms of fairness.

The relevance to the tax compliance decision covers various aspects. First, perceived peer-to-peer fairness which follows a logic similar to: "if others (don't) pay their taxes then I am (not) going to pay them as well". Second, the tax system itself might be evaluated in terms of fairness before taking a compliance decision. The items under scrutiny could be the magnitude of the fine, the frequency and modality of performed audits, the tax progressivity and thus the level of tax equity with respect to one's income. Finally, taxes are levied in order to finance public projects which may lead to an evaluation of the personal gain from paying taxes and receiving public good consumption in return, or also of the efficiency of public spending.

One way to test the relevance of social norms is to conduct cross-country surveys. By comparing responses from different cultures with similar fiscal systems, different tax attitudes and compliance rates emerge. Alm et al. (1995) provide a rough summary of the main findings of such studies. Drivers of tax compliance can be classified into moral (compliers view tax evasion as immoral and "moral appeal" tends to have positive effects on compliance), reputational (low social standing of tax evaders), peer effects (friends of tax evaders tend to evade more), perception of fairness, trust, and social cohesion. Alm et al. (1995) conducted a tax compliance experiment in the laboratory with Spanish subjects replicating an earlier study that was run in the United States. From the comparison of the two studies, it emerges that the Spanish subjects tend to comply less than their American counterparts in absolute terms. In the absence of a public good, with a fine rate of 2, a 30 % tax rate, and a 5 % probability of being audited, compliance of the American subjects is 27 % on average, whereas in Spain it amounts to only 7 %. However, the Spaniards turn out to be much more sensitive toward fiscal policies, such as changes in the tax rate, the audit rate, or the magnitude of the fine. While reminding the reader that the only difference between the two experiments lies in the cultural origin of the subject pools, the authors conclude that the social norm of compliance, which can also be defined as "tax morality", might be the reason for the difference in responses. It also emerges that there are different types of taxpayers: those who always comply and those who never comply, utility maximizers, subjects that behave according to prospect theory overweighting low probabilities, highly policy-sensitive subjects, and some who are at times cooperative and at times free-riders.

The power of social norms is also determined by the interaction and enforcement among individuals. The relevance of peer effects is bolstered by Vogel's analysis of a survey with Swedish respondents (Vogel 1974), which shows that contacts with tax evaders decreases tax compliance, weakening the social norm of compliance and, as a consequence, also the stigma of evading. On the other hand, conforming to tax-paying peers might also yield a positive return, based on the individual's level of intrinsic tax morale.

A formal model, which is still embedded in the EUT framework, was developed by Myles and Naylor (1996). Conformity to social groups yields an additional payoff to the taxpayer, which depends on the size of the group itself. Moreover, the non-evasion equilibrium could potentially be turned over by small changes in the tax rate, leading to an evasion epidemic with tipping point behavior.

Tightly linked to the concept of social norms and group conformity are the models considering the psychic costs of evasion, as described by Gordon (1989). The evaded amount becomes a function of an additional parameter which is determined by the personal level of morale and peer effects. The positive relationship between tax morale and tax compliance has been tested and confirmed numerous times in the laboratory (for an extensive survey and discussion of experimental results see Torgler 2002). In a separate article, Torgler (2003) shows empirically that the level of tax morale itself is influenced by formal and informal institutions, such as direct participation rights and trust in the government. He defines tax morale as "the intrinsic motivation to pay taxes" or "the willingness to pay taxes by the individuals."

We have seen that the tax-compliance decision is not only determined by the absolute levels of the classic parameters, which are the fine rate, the tax rate, and the probability of being caught in a random audit. Psychological and cultural factors also play a role, as well as peer interactions. The latter exhibit imitative patterns based on lagged events and give rise to the need for a dynamic modeling approach. In the following section we will present the tool of agent-based modeling, which represents a way to bridle the rise in complexity of stylized facts that potentially influence the tax-compliance decision.

5.6 From Top Down to Bottom Up—From Experimental Economics to an Experimental Economy

The ultimate purpose of research on tax evasion is undoubtedly to find policies able to increase tax compliance and hence the overall tax levy. A number of interplays and complexities characterizing the system (e.g., country) under analysis have to be considered in order to fit the outcome-predicting model as closely as possible to the underlying reality without too much loss of generality.

It is, however, a challenging task and not always possible to disentangle the behavioral and economic elements affecting the tax-compliance decision by analyzing the available empirical data. Parameter values needed for a correct calibration could be obtained, for example, by performing field studies and conducting laboratory experiments. The idea of using experimental data (but not solely) in order to feed a computer-simulated replica of society was suggested by Duffy (2006), and even though calibration with experimental data has not yet become a widespread habit, we will see one agent-based model that put these suggestions into practice.

Agent-based models offer an alternative approach to deal with complexity, and the available tools make it possible to take interactions and heterogeneities into account without necessarily abandoning the simplicity of representation. It is possible to artificially recreate an experimental economy that reacts according to the model we choose. The parameter values can be calibrated according to experimental or empirical findings; and, most importantly, this technique allows simulation with heterogeneous agents and social interactions in a dynamic environment. The results of agent-based simulations are then compared with real data, allowing for a more detailed understanding of the underlying social and behavioral dynamics.

5.6.1 Group Conformity and Social Norms

Bloomquist (2006) reviewed three agent-based simulation models applied to an environment of tax compliance, namely, Mittone and Patelli (2000), Davis et al. (2003), and Bloomquist (2004).

Mittone and Patelli (2000) use the model of Myles and Naylor (1996) as a basis, which considers group conformity and the social norm of tax compliance. Psychological costs are also included in the model, as originally proposed by Gordon (1989), but without making them depend on the evaded amount. The underlying idea is that, no matter how much income is hidden, once the decision to evade is put into practice, the "honest citizen" status is lost.

Agent heterogeneity is captured by introducing three types of subjects: the honest taxpayer, the imitative taxpayer, and the perfect free-rider, each with his or her own specific utility function. Honest agents achieve positive marginal utility effects from conforming to social rules; free-riders will contribute as little as possible; and imitative taxpayers will use the population mean of compliance as a benchmark, which is also in a way in line with the findings on peer effects described by Vogel (1974) and the findings of Porcano (1988), that the perception of existing evasion has a positive and significant effect on the own level of evasion. Additional utility gains are obtained from the introduction of a public good that depends on the amount of tax levy, as considered by the theoretical model of Cowell and Gordon (1988).

Finally, the behavioral characteristics of single agents are not static, but subject to a stochastic updating process, a genetic procedure where probabilities of type-survival are calculated based on individual utility gains over total population utility gains.

Decision rules on how much to declare follow a learning mechanism with choices being updated based on the success or failure of past compliance decisions.

The random element in the behavioral switching algorithm, in combination with the feature of imitative behavior, triggers a cycle that allows for the model to react to audit rates that are close to zero, pushing compliance to a near-zero level, even if all agents were initially honest.

On the other hand, when audits are introduced, only honest taxpayers remain after a certain number of rounds. It is interesting to notice that the type of audit procedure, uniform versus those aimed at the lower tail of contributions, does not change this result.

Another model considering peer-oriented behavior is the Multi-Agent Based Simulation (MABS) developed by Davis et al. (2003) using the software *Mathematica*. The behavioral classification of taxpayers is similar to the previous model: there are honest, susceptible (to others' behavior), and evading agents.

Initially, there are two randomly assigned types: honest and evading. An honest taxpayer might, depending on the "infection rate," become a susceptible one in case a (randomly chosen) acquaintance happens to be an evader. Susceptible agents form beliefs about the severity of the tax enforcement regime by observing the mean audit frequency of their peers in the previous period. They become evaders if this perceived severity is below a certain threshold. Finally, evaders become honest taxpayers if the belief about the severity of the audit regime is above a certain threshold or if they observe a social norm of tax compliance. The existence of a social norm of tax compliance is confirmed by having a certain number of honest taxpayers among the own acquaintances.

Also, evading agents become honest after a tax audit. The authors follow the literature on availability (Tversky and Kahneman 1973) and vividness, assuming that the subjectively determined probability of being audited is judged to be higher after a recent audit experience. This assumption, however, stands in contrast with the experimental findings of Mittone (2006) and Maciejovsky et al. (2007), who observed negative post-audit responses, which are likely to be due to the so-called Bomb-Crater Effect, according to which most recent events are judged to be unlikely to occur again immediately. A second cause is a loss-repair mechanism as described by Maciejovsky et al. (2007) and suggested also by Andreoni et al. (1998) where the fined taxpayer tries to recover the sum by evading more in the subsequent round.

By manipulating the starting proportion of evading and honest agents and letting the systems evolve over 2,000 rounds, Davis et al. (2003) find that changes in the audit rates from 0.002 to 0.030 in steps of 0.002 lead to "tipping point" behavior with abrupt changes in compliance equilibria. In all set-ups societies converge to total honesty at an audit rate of only 0.03. Although the latter result is not supported by empirical findings, it is still notable that the audit rate may be used as a device to prevent a non-compliance epidemic from happening.

The authors suggest establishing a similar experimental environment with social norms and group conformity in order to confirm the robustness of their finding.

5.6.2 The Tax Compliance Simulator (TCS)—Playing with Complexity

Bloomquist (2004) uses the software NetLogo in order to simulate a more complex agent-based environment. His Tax Compliance Simulator (TCS) is capable of testing the effects of variations in audit rates, fine rates, income visibility (wages and salaries versus other sources of income), auditor efficacy, and audit celerity after the event of an evasion. What is measured are direct effects of audits given by the additional levy, indirect effects of audits experienced by peers, and post-audit responses.

The interface of the TCS is also quite intuitive for a non-technical user and was illustrated in Bloomquist (2006, p. 423). It is composed by various sliders (such as tax rate, audit rate, penalty, etc.) through which it is possible to manipulate the desired parameter values before starting the simulation.

Diagrams at the top show the evolution of the variables of interest, as, for example, the amount of reported tax over time, whereas the distribution dynamic of full evaders, partial compliers, and full compliers appears in the window containing "turtles," the NetLogo labeling of what we called agents.

The TCS features both overweighting of low audit probabilities and overestimation. Taxpayers tend to overweight low probabilities of audit, as predicted by prospect theory (Kahneman and Tversky 1979) and experimentally assessed by Alm et al. (1992). All probabilities were given, but there was still evidence of people typically overweighting these probabilities. Overestimation, on the other hand, is a bias which is a less numerical and more psychological phenomenon: it may depend on the subjects' perception of the auditing mechanism and also past audit experiences, as hypothesized in Kastlunger et al. (2009).

In order to help account for the opportunities to evade, it is possible to determine the percentage of visible income. Visible income being subject to third-party information reporting, such as salaries, is assumed to be entirely declared, which is in line with empirical data. However, once the agent does not declare her full income it is not always true that when an audit is performed, the full evasion is detected. To account for such partial detection, a detection rate is introduced and the cost component of the decision to evade is modified accordingly.

Finally, also borrowing constraints which could incentivize lower compliance and discount rates for delayed detection with respect to the evasion event itself are taken into account.

Unlike the first two models, the TCS uses actual empirical evidence from IRS audit data when calibrating the parameters of the MABS model in order to achieve an outcome which is as close as possible to observed levels of compliance.

5.6.3 Experimental Economics—Calibration with Experimental Data

In an agent-based exercise simulating tax compliance behavior of small-business owners Bloomquist (2011) employs a relatively simple evolutionary game-theoretic approach and uses experimental data in order to calibrate the model programmed in NetLogo. Taxpayers this time are of four different types: Honest, Strategic, Defiant, and Random. Behavior of small-business owners is hypothesized to be similar to that of laboratory subjects and in order to prove this claim the author relies on third-party experimental results (Alm and Mckee 2006; Alm et al. 2008). In these tax-compliance experiments subjects were tested for various behavioral mechanisms, but for comparability reasons only data of the "no treatment" subject pools was extrapolated for the calibration purpose at issue. Compliance rate histograms clearly exhibit the typical bimodal distribution around zero and one, hence confirming the existence of honest and defiant taxpayers.

Initially the combination of behavioral taxpayer types is selected such that laboratory results in terms of mean and mode are matched as closely as possible. In this first run no neighborhood effects were included because laboratory subjects were not able to see what others were doing. From the comparison of the simulation results for five different audit rates from 0 to 0.40 with and without risk aversion,[5] it can be inferred that the inclusion of risk aversion with probability weighting gives more precise results for audit rates ranging from 0 to 0.10, confirming the findings of Bernasconi (1998).

As a second step four different scenarios[6] of agent-based modeling were matched to real data gathered from the IRS National Research Program (NRP) study. The subsample of individuals with income stemming only from Sole Proprietorship (Schedule C—Profit or Loss From Business) was used and cases of overcompliance were normalized to full compliance. None of these four simulations was able to match the average NRP compliance rate at which point the agent group defined as "random taxpayers" was excluded from the runs, assuming that small-business owners exhibit less random behavior than the students who participated in the lab.

The best match with the new setup was found in absence of neighborhood effects, which is also clear from the comparison of the histogram of compliance rates of the simulation with real data excluding neighborhood effects from playing a dominant role.

5.6.4 *A Model of Citizenship*

Pellizzari and Rizzi (2014) developed an agent-based model of tax compliance that contemplates two types of agents: taxpayers and the government. Taxpayers maximize their utility based on net income and also considering the perceived level of public expenditure. Again, the role of the public good is considered as being relevant to the tax-compliance decision, but in a more sophisticated and dynamic way than we already saw in Mittone and Patelli (2000). In addition, an array of individual characteristics are included—namely, risk aversion; relative preference for public expenditure; an innate tendency to pay taxes, which we could define as intrinsic

[5] Risk aversion in this case describes the mechanism of overweighting small probabilities of being audited as described (also in Bernasconi, 1998).

[6] Risk aversion versus no risk aversion and neighborhood effects versus no neighborhood effects.

tax morale; and, as in Mittone and Patelli (2000), group conformity modeled as the expectation about other agents' compliance behavior. Audit probabilities are not exactly known and inferred subjectively by observing audit dynamics among peers.

Moreover, the authors propose three different institutional frameworks based on the power a government is able to exert (high, average, and low). The taxpayers are divided into three types based on their level of "citizenship," which is defined as a combination of the preference for public expenditure, group conformity, and tax morale.

Under a weak enforcement regime and with a low level of citizenship the authors observed high levels of tax evasion. Even with high levels of enforcement, full compliance cannot be achieved. Moreover, *ceteris paribus*, the higher the tax rate, the lower the compliance level of their agents. This finding is in contrast with the predictions of the theoretical EUT model developed by Yitzhaki (1974) but confirms a large body of empirical evidence.

Government power still plays a significant role in societies with an average level of citizenship, even though elasticities decrease with respect to low levels of citizenship. Finally, the role of government almost disappears for high levels of citizenship, where tax evasion approaches near-zero levels regardless of the enforcement regime.

Overall, citizenship is found to have a larger marginal effect on compliance than enforcement power by the government, but both concepts are necessary to enhance the level of tax compliance in a society.

To take the findings further, it could also be interesting to read these results in light of the somewhat contrasting prescriptions coming from the literature on motivation crowding-out, surveyed for example by Frey and Jegen (2001): the intrinsic motivation to pay taxes could therefore be decreased by the mere presence of an extrinsic mechanism of punishment.

5.7 Conclusions

The aim of this survey was to take the reader on a tour through the very rich body of existing literature on tax evasion. Not only the plurality of determinants of tax compliance, but also the multiplicity of methodologies and analytical approaches make it a challenging task to provide policy makers with useful indications.

We started from the classical models of tax evasion which represent an ideal starting point for more realistic considerations and modifications. Thereafter, institutional, social and psychological factors, among others, have been found to be highly relevant when a taxpayer decides how much to declare to the tax authority. Also, survey data and laboratory experiments have shown that interaction among taxpayers in the form of peer effects and psychic costs of evasion cannot be disregarded.

As the famous physicist Stephen Hawking so wisely predicted, "the next century will be the century of complexity." We now face the moment in which we have to tame and coordinate all these different elements in order to maintain the informative value of new findings.

To do so, the *ceteris paribus* approach and the search for equilibria should be relaxed, giving space to a new notion of *ceteris mutabilibus*[7] and asymptotic dynamics. Innovative tools are needed, and the impressive computing power of modern devices serves this purpose well. We presented some examples of agent-based models to provide the reader with food for thought by highlighting their flexibility in accounting for both heterogeneous agents as well as different kinds of parameters, in addition to the possibility for interaction and the dynamic nature of such simulations. Additionally, individual characteristics can be matched with evidence stemming from human subjects tested in the lab, as we have seen in Bloomquist (2011).

The future focus of research on tax compliance should hence be on continuing the multidisciplinary approach in determining the drivers of tax evasion, on one hand, while properly administering, elaborating and integrating old and new findings, on the other, with the final aim being to provide policymakers with ever-improving policy advice on how to increase the overall level of tax compliance.

References

Allingham, M. G., & Sandmo, A. (1972). Income tax evasion: A theoretical analysis. *Journal of Public Economics, 1*(3–4), 323–338.

Alm, J., & McKee, M. (2006). Audit certainty, audit productivity, and taxpayer compliance. *Andrew Young School of Policy Studies Research Paper*, 06–43.

Alm, J., McClelland, G. H., & Schulze, W. D. (1992a). Why do people pay taxes? *Journal of Public Economics, 48*, 21–38.

Alm, J., Jackson, B., & McKee, M. (1992b). Estimating the determinants of taxpayer compliance with experimental data. *National Tax Journal, 45*, 107–114.

Alm, J., Sanchez, I., & Juan, de A. (1995). Economic and noneconomic factors in tax compliance. *Kyklos, 48*, 3–18.

Alm, J., Deskins, J., & McKee, M. (2008). Do individuals comply on income not reported by their employer? *Public Finance Review, 37*, 120–141.

Alm, J., Bloomquist, K., & McKee, M. (2010). On the external validity of tax compliance experiments. Paper presented at the Annual Meeting of the National Tax Association, November 2010.

Andreoni, J., Erard, B., & Feinstein, J. (1998). Tax compliance. *Journal of Economic Literature, 36*(2), 818–860.

Arrow, K. (1965). *Aspects of the theory of risk bearing*. Helsinki: Yrjo Jahnsson Saatio.

Becker, G. (1974). Crime and punishment: An economic approach. *Essays in the Economics of Crime and Punishment, I*, 1–54.

Bernasconi, M. (1998). Tax evasion and orders of risk aversion. *Journal of Public Economics, 67*, 123–134.

Bernasconi, M., Corazzini, L., & Seri, R. (2014). Reference dependent preferences, hedonic adaptation and tax evasion: Does the tax burden matter? *Journal of Economic Psychology, 40*(0), 103–118.

[7] "All else changeable."

Bloomquist, K. M. (2004). Modeling taxpayers' response to compliance improvement alternatives. *Annual conference of the North American association for computational social and organizational sciences*, Pittsburgh, PA.
Bloomquist, K. M. (2006). A comparison of agent-based models of income tax evasion. *Social Science Computer Review, 24*(4), 411–425.
Bloomquist, K. M. (2011). Tax compliance as an evolutionary coordination game: An agent-based approach. *Public Finance Review, 39*, 25–49.
Buchanan, J. M. (1987). The constitution of economic policy. *The American Economic Review, 77*(3), 243–250.
Clotfelter, C. (1983). Tax evasion and tax rates: An analysis of individual returns. *Review of Economics and Statistics, 65*(3), 363–373.
Conlisk, J. (1996). Why bounded rationality? *Journal of Economic Literature, 34*(2), 669–700.
Cowell, F. A., & Gordon, J. P. F. (1988). Unwillingness to pay—tax evasion and public good provision. *Journal of Public Economics, 36*, 305–321.
Davis, J. S., Hecht, G., & Perkins, J. D. (2003). Social behaviors, enforcement, and compliance dynamics. *The Accounting Review, 78*(1), 39–69.
Dhami, S., & al-Nowaihi, A. (2007). Why do people pay taxes? Prospect theory versus expected utility theory. *Journal of Economic Behavior and Organization, 64*(1), 171–192.
Duffy, J. (2006). Agent-based models and human subject experiments. In L. Tesfatsion & K. L. Judd (Eds.), *Handbook of Computational Economics* (vol. 2, pp. 949–1011). North-Holland: Elsevier.
Elster, J. (1989). Social norms and economic theory. *The Journal of Economic Perspectives, 3*(4), 99–117.
Frey, B. S., & Feld, L. P. (2002). Deterrence and morale in taxation: An empirical analysis. *CESifo Working Paper No. 760*, Munich.
Frey, B. S., & Jegen, R. (2001). Motivation crowding theory. *Journal of Economic Survey, 15*(5), 589–611.
Friedland, N. (1982). A note on tax evasion as a function of the quality of information about the magnitude and credibility of threatened fines: Some preliminary research. *Journal of Applied Social Psychology, 12*(1), 54–59.
Gordon, J. P. F. (1989). Individual morality and reputation costs as deterrents to tax evasion. *European Economic Review, 33*, 797–805.
Graetz, M. J., & Wilde, L. L. (1985). The economics of tax compliance: Facts and fantasy. *National Tax Journal, 38*, 355–363.
Guala, F., & Mittone, L. (2005). Experiments in economics: External validity and the robustness of phenomena. *Journal of Economic Methodology, 12*, 495–515.
Kahneman, D. (2003). Maps of bounded rationality: Psychology for behavioral economics. *The American Economic Review, 93*(5), 1449–1475.
Kahneman, D., & Tversky, A. (1979). Prospect theory: An analysis of decision under risk. *Econometrica, 47*(2), 263–292.
Kastlunger, B., Kirchler, E., Mittone, L., & Pitters, J. (2009). Sequences of audits, tax compliance, and taxpaying strategies. *Journal of Economic Psychology, 30*(3), 405–418.
Maciejovsky, B., Kirchler, E., & Schwarzenberger, H. (2007). Misperception of chance and loss repair: On the dynamics of tax compliance. *Journal of Economic Psychology, 28*(6), 678–691.
Mittone, L. (2006). Dynamic behaviour in tax evasion: An experimental approach. *Journal of Socio-Economics, 35*(5), 813–835.
Mittone, L., & Patelli, P. (2000). Imitative behaviour in tax evasion. In F. Luna & B. Stefansson (Eds.), *Economic simulations in swarm: Agent-based modelling and object oriented programming SE—5* (Vol. 14, pp. 133–158). Springer: New York.
Myles, G. D., & Naylor, R. A. (1996). A model of tax evasion with group conformity and social customs. *European Journal of Political Economy, 12*, 49–66.
Pellizzari, P., & Rizzi, D. (2014). Citizenship and power in an agent-based model of tax compliance with public expenditure. *Journal of Economic Psychology, 40*, 35–48.
Porcano, T. M. (1988). Correlates of tax evasion. *Journal of Economic Psychology, 9*, 47–67.

Pratt, J. W. (1964). Risk aversion in the small and in the large. *Econometrica, 32*(1), 122–136.
Reyna, V. F., & Brainerd, C. J. (2008). Numeracy, ratio bias, and denominator neglect in judgments of risk and probability. *Learning and Individual Differences, 18*(1), 89–107.
Selten, R. (2002). What is bounded rationality? In G. Gigerenzer & R. Selten (Eds.), *Bounded rationality: The adaptive toolbox* (pp. 13–36). MA: MIT Press.
Simon, H. A. (1972). Theories of bounded rationality. In C. B. McGuire & R. Radner (Eds.), *Decision and organization: A volume in honor of Jacob Marschak* (pp. 161–176). Amsterdam: North Holland.
Simon, H. A. (1982). *Models of bounded rationality* (Vol. 2). Cambridge: MIT Press.
Slemrod, J., & Yitzhaki, S. (2002) Tax avoidance, evasion, and administration. *Handbook of Public Economics, 3*, 1423–1470.
Torgler, B. (2002). Speaking to theorists and searching for facts: Tax morale and tax compliance in experiments. *Journal of Economic Surveys, 16*(5), 657–683.
Torgler, B. (2003). Tax morale and institutions. *CREMA Working Paper Series* 2003–9, Basel.
Tversky, A., & Kahneman, D. (1973). Availability: A heuristic for judging frequency and probability. *Cognitive Psychology, 5*, 207–232.
Tversky, A., & Kahneman, D. (1992). Advances in prospect-theory—cumulative representation of uncertainty. *Journal of Risk and Uncertainty, 5*, 297–323.
Vogel, J. (1974). Taxation and public opinion in Sweden: An interpretation of recent survey data. *National Tax Journal, 27*(4), 499–514.
Yitzhaki, S. (1974). A note on income tax evasion: A theoretical analysis. *Journal of Public Economics, 3*, 201–202.

Chapter 6
Exploring Reputation-Based Cooperation:

Reputation-Based Partner Selection and Network Topology Support the Emergence of Cooperation in Groups

Daniele Vilone, Francesca Giardini and Mario Paolucci

6.1 Introduction

> Two neighbors may agree to drain a meadow, which they possess in common because it is easy for them to know each other's mind; and each must perceive, that the immediate consequence of his failing in his part, is the abandoning of the whole project. But it is very difficult, and indeed impossible, that a 1000 persons should agree in any such action; it being difficult for them to concert so complicated a design, and still more difficult for them to execute it; while each seeks a pretext to free himself of the trouble and expense, and would lay the whole burden on the others. (David Hume, A Treatise of Human Nature p. 239)

Humans show levels of cooperation among non-kin that are unparalleled among other species. This difference becomes striking when facing social dilemmas, i.e., situations in which cooperation is hard to achieve because the best move for an individual does not produce the best outcome for the group. Public goods games (PGG) represent a clear exemplification of this conflict between individual incentives and social welfare. If everybody contributes to the public good, cooperation is the social optimum, but free-riding on others' contributions represent the most rewarding option at the individual level.

If norms, conventions and societal regulations have been proven effective in preventing the collapse of public goods (for a review, see (Ostrom 1990; Ostrom 2005)), when individuals are faced with unknown strangers, with little or no opportunities for future re-encounters, cooperation easily collapses, unless punishment for non-cooperators is provided (Fehr and Gachter 2000). An alternative solution is represented by reputation, that allows to identify and avoid cheaters (Nowak and

D. Vilone (✉) · F. Giardini · M. Paolucci
Laboratory of Agent-Based Social Simulation, Institute of Cognitive Sciences and Technologies – CNR, Via Palestro 32, Rome 00185, Italy
e-mail: daniele.vilone@gmail.com

F. Giardini
e-mail: francesca.giardini@istc.cnr.it

M. Paolucci
e-mail: mario.paolucci@istc.cnr.it

F. Cecconi (ed.), *New Frontiers in the Study of Social Phenomena*,
DOI 10.1007/978-3-319-23938-5_6

Sigmund 1998b; Giardini and Conte 2012). Indirect reciprocity supported by reputation (Alexander 1987) can be one of the mechanisms explaining the evolution of cooperation in humans (Milinski et al. 2002), especially in large groups of unrelated strangers who can, through language, actively communicate about their past experiences with cheaters (Smith 2010).

As such, the exchange of evaluative information about other agents, i.e., gossip, may effectively bypass the "second-order free-rider problem", wherein the costs associated with solving one social dilemma produces a new one (Hardin 1968; Kiyonari and Barclay). This is the case of punishment: cooperators who do not sustain the costs of punishment are better off than cooperators who also punish. Punishment as a solution to the dilemma of cooperation entails another social dilemma, because punishment is costly and cooperators who do not punish are better off than punishers. On the other hand, costless gossip should not imply such a second-order free-rider problem. In addition to costly punishment and reputation, ostracism may represent a third solution against free-riders. However, the direct effect of ostracizing someone is that group size decreases, thus automatically reducing maximal contribution levels to the public good for all remaining periods. Maier-Rigaud and colleagues show that in laboratory experiments, PGG with ostracism opportunities increases contribution levels and, unlike monetary punishment, also has a significant positive effect on net earnings (Maier-Rigaud et al. 2005).

Models of indirect reciprocity usually take into account dyadic interactions (Nowak and Sigmund 1998b), or group interactions in a mutual aid game (Panchanathan and Boyd 2004), in which providing help has a cost for the helper but it also increases his/her image score, i.e., a publicly visible record of the individual's reputation. Image score increases or decreases according to individuals' past behaviors, thus providing a reliable way to discriminate between cheaters and cooperative players. Both in computer simulations (Nowak and Sigmund 1998b), and in laboratory experiments with humans (Wedekind and Milinski 2000), cooperation can emerge and be maintained through image score.

When individuals facing a social dilemma can know other players' image score, cooperation can emerge in small groups, as showed by Suzuki and Akiyama (Suzuki and Akiyama 2005). In their work, cooperation can emerge and be maintained in groups of four individuals; though, when group size increases, there is a concomitant decrease in the frequency of cooperation. The authors explain this decline as due to the difficulty of observing reputations of many individuals in large communities. This can be true of unstructured communities, but this rarely happens in human societies, characterized by interaction networks.

To account for the role of social structure, we designed a PGG in which players' interactions depend on the kind of network and on the possibility of actively choosing a subset of group members. More specifically, we compare cooperation levels among agents placed on a small-world network (Watts and Strogatz 1998), defined by short average path lengths and high clustering, to the performance of agents on a bi-partite graph (Diestel 1997; Gómez-Gardeñes 2011). The latter is generally used to model relations between two different classes of objects, like affiliation networks linking members and the groups they belong to. This structure is especially interesting for us because it is especially suited for partner selection, as it happens when a club refuses membership to a potential associate.

Here, we are interested in exploring the effect of network structure on the emergence of cooperation in a PGG. We compare two different network topologies and we show that reputation-based partner choice on a bi-partite graph can make cooperation thrive also in large groups of agents. We also show that this effect is robust to number of generations, group size and total number of agents in the system.

6.2 The Computational Model

We consider a population of N individuals. In each round of the game, g agents are picked up at random to play a PGG among themselves. Players can cooperate contributing with a cost c to a common pot, or can defect without paying anything. Then, the total amount collected in the pot is multiplied for a benefit b and equally distributed among all the group members, without taking into account individual contributions. At the end of each interaction, being X the number of contributors in the group, cooperators' payoffs equals $(Xb/g-1)c$, whereas defectors' payoffs is Xbc/g. At the collective level, the best outcome is achieved when everyone cooperates, but cheaters are better off, because defection permits to avoid paying any costs when the number of cooperators is lower than gc/b.

Among the many solutions offered (Fehr and Gachter 2000), Suzuki and Akiyama (Suzuki and Akiyama 2005) design a modified PGG in which agents can identify cheaters thanks to the so-called image score (Nowak and Sigmund 1998b; Nowak and Sigmund 1998a).

The basic features of our model are the same of the one by Suzuki and Akiyama: in particular, each player i is characterized by two integer variables: the image score $is \in [-Smax, Smax]$ and the strategy $ki \in [-Smax, Smax+1]$, being $Smax \geq 0$ a parameter of the model. When selected to play a round of the game, an individual cooperates if the average image score $\langle s \rangle g$ of its opponents is equal to or higher than its own strategy ki, otherwise it will not contribute. At the end of the round, the image score of the player is increased by one in case of cooperation, otherwise it is decreased by the same quantity. In any case, si remains in the allowed interval $[-Smax, Smax]$: if an agent has an image score of $Smax$ $(-Smax)$ and contributes, nothing happens to its image score. At the initial stage, all the image scores and fitness levels are set to zero, whilst the strategies are randomly distributed among the individuals.

The image score is intended to give a quantitative evaluation of the public reputation of an individual in the scope of indirect reciprocity: if contributing once is rewarded by future contributions by other individuals, then any cooperative act must be recognized and considered positively by the entire population; on the other hand, the variability of the strategies describes the different attitudes and expectations of the single agents (Nowak and Sigmund 1998b).

After m rounds, reproduction takes place. Again, we apply the same evolutionary algorithm used by Suzuki and Akiyama (Suzuki and Akiyama 2005). For N times we select at random a pair of individuals and with probability P we create a new individual inheriting the strategy of the parent with the highest fitness. Then parents are put again in the population, and offspring is stored in another pool. When this selection process has happened N times, the old population is deleted and replaced with the offspring. It is worth noticing that offspring inherit only the parent's strategy, while their image score and fitness is set equal to zero. Finally, we repeat this sequence (m rounds followed by the reproduction stage), for an adequate number of generations. The simulation lasts until the system reaches a final (steady or frozen) configuration.

For sake of clarity, we observe that strategies defined as ($k \leq 0$) are the more "cooperation prone", with the limit case of $k = -Smax$ which is an absolute cooperator, while the positive ones are the "cooperation averse" strategies, with the limit case of $k = Smax + 1$ representing an inflexible defector.

Moving from the model described above, we are interested in testing whether two different network structures can promote cooperation for different group size and what effect partner selection can have in such an environment.

6.3 Results

Suzuki and Akiyama tested their model for a given set of parameters with the following values: total population of 200 agents ($N = 200$), cost of cooperation set to 1 ($c = 1$), for a benefit of 0.85 multiplied by the size of the group ($b = 0.85$ g). The highest possible value of the image score equals 5 (Smax = 5), and the total number of rounds of the game is set to 800 ($m = 800$). Their results show that a cooperative strategy can evolve and invade a population when group size is small, but it does not survive when groups are large. For medium-sized communities, a coexistence between cooperators and defectors is possible.

The first step of the study present in this paper is a check of the robustness of Suzuki and Akiyama results with respect to the values of the model parameters. A check of the role of m and N is reported already in (Suzuki and Akiyama 2005): it is claimed that the outcome is not relevantly influenced by the value of these two quantities, so we focus here on b, P and $Smax$.

The role of b in the PGG is quite clear in literature. Normally it is set to 3 independently from the group size (Walker and Williams 1994). Using this value, we found that the final cooperation level decreases sharpenly as g increases, as shown in Fig. 6.1.

The fact that in Suzuki's and Akiyama's work such decreasing is much slower is due to the fact that being b proportional to the group size, the number of contributors needed to make cooperation advantageous remains constant in g instead of decreasing with it. On the other hand, even though less dramatic, the decrease is anyway observed, indicating that the negative effect of large groups on cooperation is stronger and it might depend on the PGG dynamics.

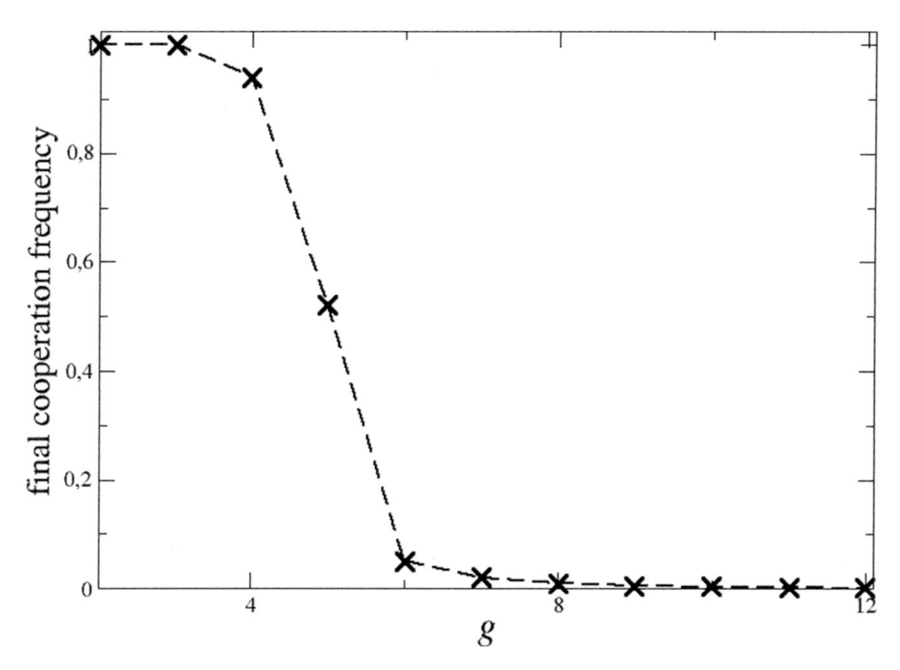

Fig. 6.1 Behaviour of the final frequency of cooperative actions as a function of the group size *g*. All the parameters are the same of the paper by Suzuki and Akiyama (Suzuki and Akiyama 2005), except $b = 3$. Each point averaged over 1000 realizations

Concerning the behaviour of the model as a function of the parameter *P* that refers to the probability of inheriting a given strategy, we tested three different values: $P = 0.9$ as in (Suzuki and Akiyama 2005), $P = 0.75$ and $P = 1.0$. As it can be easily seen in Fig. 6.2, there is no fundamental difference due to the exact value of this parameter.

Finally, changing the value of *Smax*, we see that up to Smax ≃ 15, the behavior of the system is rather homogeneous, as shown in Fig. 6.3.

Our results show that the behavior of the model is actually robust for a large range of the parameters at stake, thus replicating Suzuky and Akiyama's results.

6.3.1 Effects of the Network Topology

In order to enlarge the scope of the model, we tested the effects of different network structures, thus introducing some adaptations of the original model. The first change we made was in the mechanism of assortment. In the original model, every player had the same probability to interact with every other agent, therefore the population is placed on a total connected graph (CG). This configuration is rather unrealistic, especially when we consider groups bigger than a given size. It is then interesting to test the model behavior over more realistic, even though still abstract, networks. The first example we take under consideration is the so-called small-world network

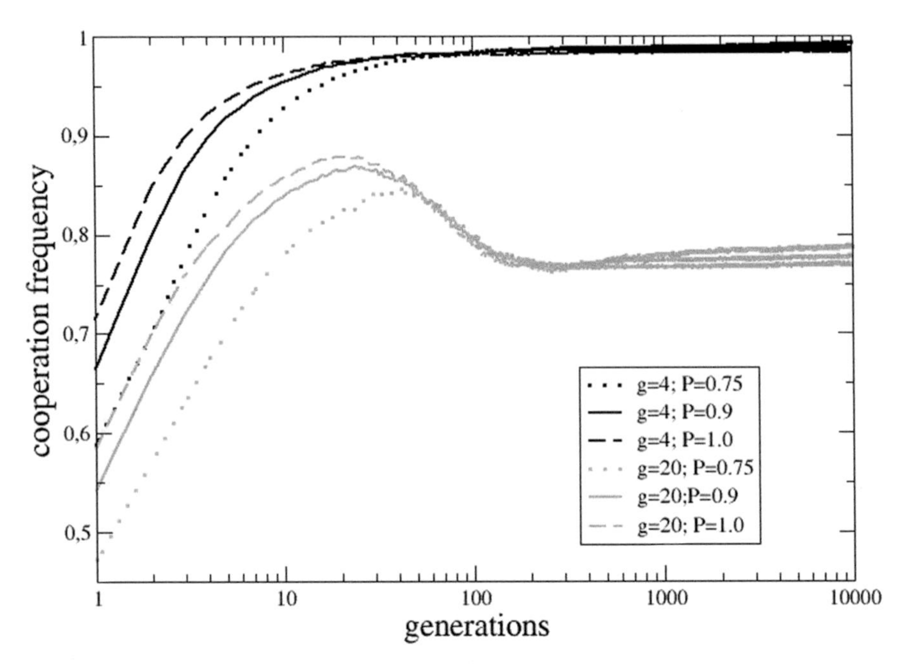

Fig. 6.2 Behaviour of the frequency of cooperative actions as a function of the number of generations for three different values of P : 0.75, 0.90 and 1.0. The remaining parameters are the same of the paper by Suzuki and Akiyama (Suzuki and Akiyama 2005). Each curve averaged over 1000 realizations

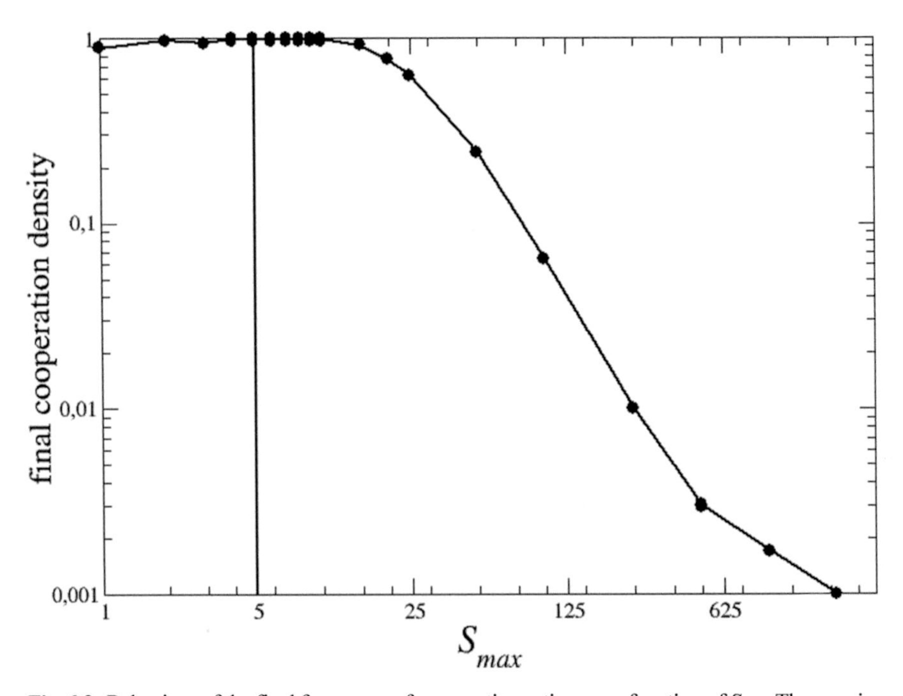

Fig. 6.3 Behaviour of the final frequency of cooperative actions as a function of S_{max}. The remaining parameters are the same of the paper by Suzuki and Akiyama (Suzuki and Akiyama 2005), the vertical line for $S_{max} = 5$ specifies the value utilized in the same reference. Each point averaged over 1000 realizations

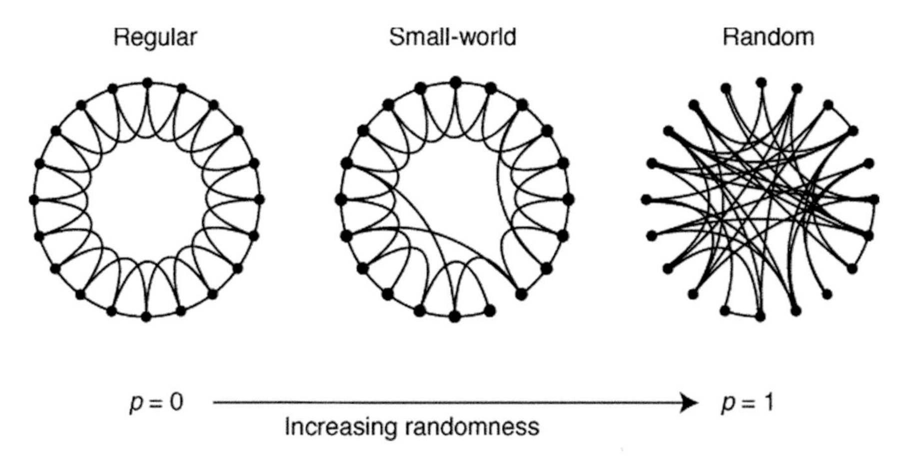

Fig. 6.4 Construction of SWN according Watts-Strogatz procedure (Watts and Strogatz 1998)

(SWN), as conceived by Watts and Strogatz in (Watts and Strogatz 1998). In short, a SWN, is a regular ring with few short-cuts linking originally far away nodes. It is constructed as shown in Fig. 6.4: we start from a ring where each node is connected with 2k nearest neighbors. Then, with probability p, each link is rewired (one of the node is left fixed, the other is changed), so that it finally leads to the creation of a network with pNk short-cuts. As shown in reference (Watts and Strogatz 1998), for $1/Nk < p < 1/10$ the network shows the typical small-world effect: even though at local level the system behaves as a regular lattice, i.e., an individual placed in a SWN cannot distinguish the network from a regular one just watching his/her neighbors (high clustering coefficient), at a global level the average distance between two randomly selected individuals is very short (proportional to the logarithm of the system size), unlike the regular network.

In order to make the model work on this topology, we had to adapt the model dynamics to the specific situation. In particular, instead of extracting g agents at each round, we picked up a single player at each round and $g-1$ of its neighbors. In order to be sure that each individual had at least $g-1$ neighbors, we set $k=g-1$. Moreover, at the end of each generation, the offspring was randomly placed on the pre-existent network, which is defined at the beginning and does not change until the end of the simulation. Anyway, averaging over different realizations, each one has its own networks, so that the averages are also over the topology.

In Fig. 6.5 and 6.6 we see the cooperation frequencies for two values of g and different sizes of the system. Basically, when group size is bigger than two, the system shows an interesting behavior: in particular, the dynamics is always driving the system towards the achievement of complete cooperation, even though the timing can vary. Full cooperation is reached more rapidly for small values of N, whilst it can take up to thousands of generations for larger systems. It is worth noticing (Cfr. Fig. 6.6), that this consensus time seems to reach its limit value already for

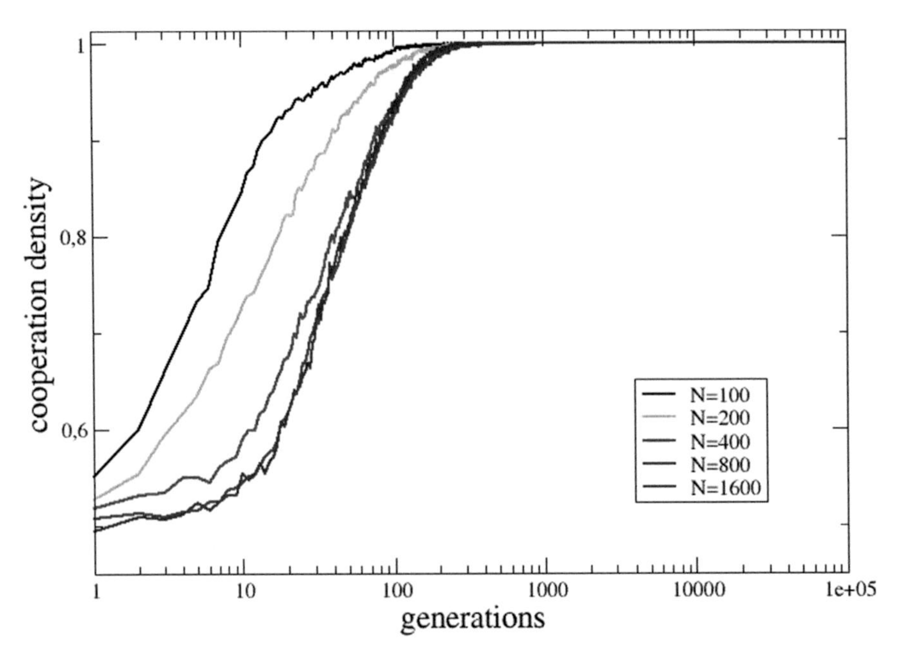

Fig. 6.5 Behaviour of the frequency of cooperative actions in a SWN with $p = 0.05$ as a function of the number of generations for $g = 2$ and different values of N (from top to bottom: 100, 200, 400, 800 and 1600). The remaining parameters are the same of the paper by Suzuki and Akiyama (Suzuki and Akiyama 2005). Each curve averaged over more than 1000 realizations

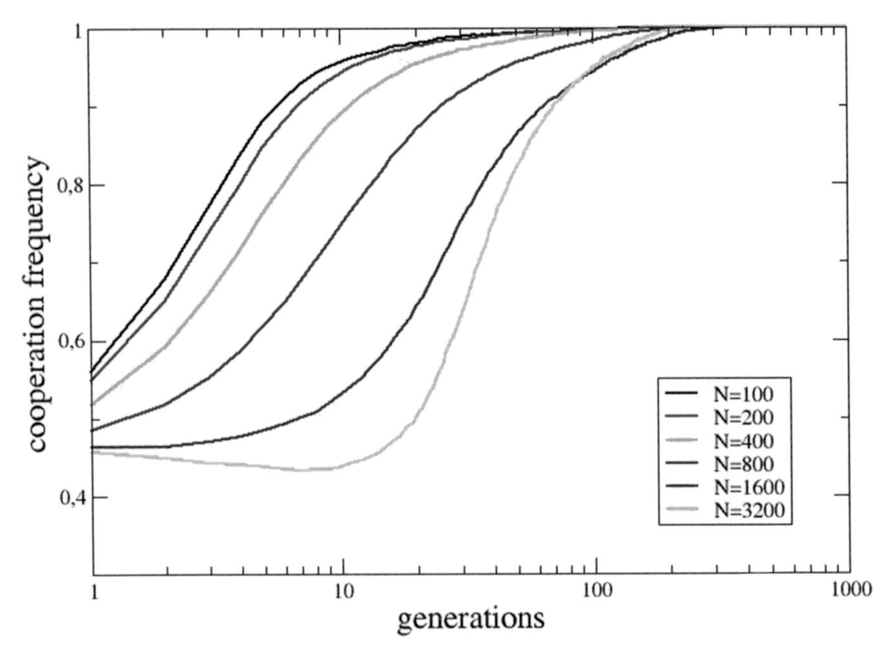

Fig. 6.6 Behaviour of the frequency of cooperative actions in a SWN with $p = 0.05$ as a function of the number of generations for $g = 4$ and different values of N (100, 400 and 1600). The remaining parameters are the same of the paper by Suzuki and Akiyama (Suzuki and Akiyama 2005). Each curve averaged over more than 1000 realizations

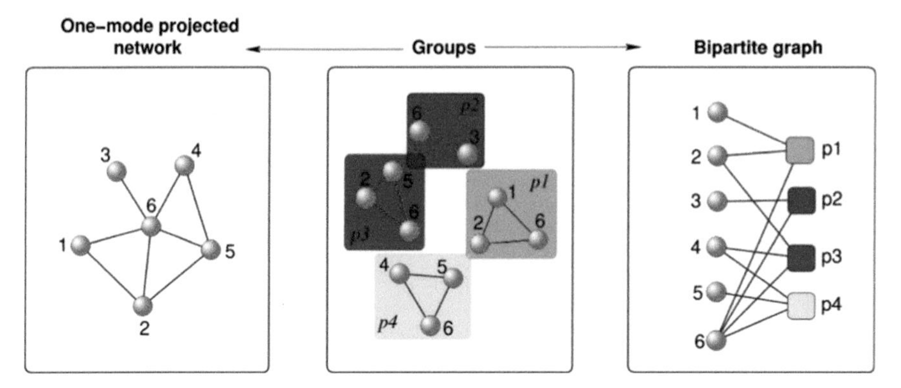

Fig. 6.7 Structure of a bipartite graph compared with a classical network (J. Gómez-Gardeñes *et al.* 2011)

$N = 3200$; in such case we can also distinguish an initial small decrease of cooperation rate before the final (steep) increase.

In short, these results demonstrate that the small-world topology in itself makes full cooperation possible, although this requires some time, as we are going to discuss in the next sections.

Another topological configuration that accounts better for the complexity of real interactions among individuals is the so-called bipartite graph (BG) (Diestel 1997; Gómez-Gardeñes 2011). A bipartite representation contains two types of nodes denoting agents and groups, respectively. It implies that connections can be established only between nodes of different types and no direct connection among individuals is allowed. Thus, such a bipartite representation preserves the information about the group structure: if two individuals belong to the same three groups, they are "more" connected than two other individuals who are members of the same group. These two pairs would be equally represented in the classical one-mode projected network, while with the bipartite graph this mesoscopic level of interactions is better depicted, as illustrated in Fig. 6.7.

Also in this case we adapted the original dynamics of the model to this kind of network. In particular, the graph has N individuals distributed into M groups, each group composed of a certain number (g) of members. At the beginning of each round, the network is built in this way: given $F \in (0, 1)$, we set gF initial members for each groups so that each individual belongs exclusively to one group. For instance, if $N = 150$, $g = 20$ and $F = 0.75$ (then $M = 10$), at this stage we would have 15 agents in the first group, other 15 in the second one and so on until the last 15 in the tenth group. Then, each group must be completed choosing five individuals from the set of those who do not belong to any group.

This can be done in two different ways:

1. by randomly picking $(1 - F)g$ agents among the rest of the population;
2. or, by selecting them according to their reputation, i.e., their image scores. In this latter case, we have a partner selection mechanism, therefore an external player is randomly selected by the group, but accepted only if its image score

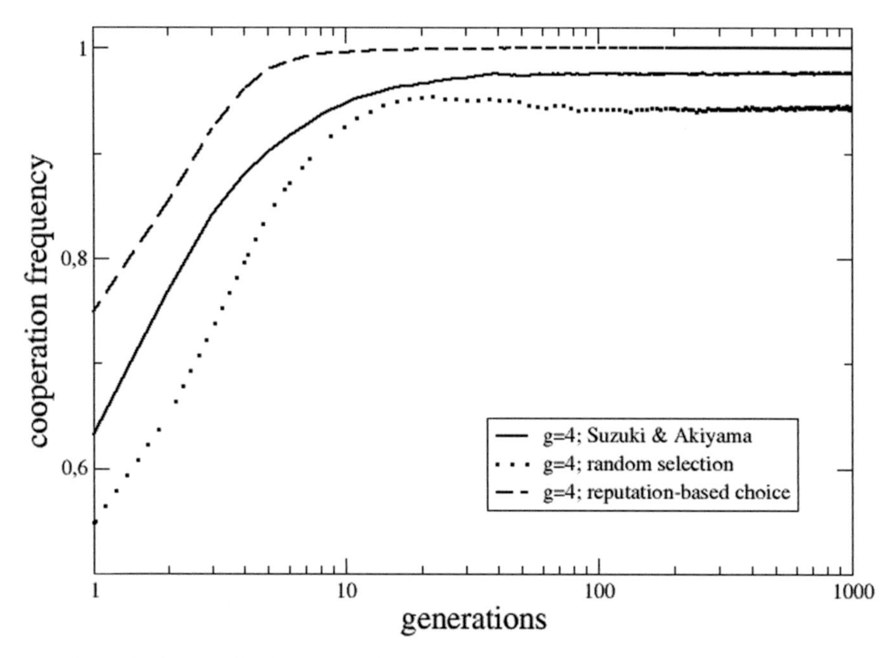

Fig. 6.8 Behaviour of the frequency of cooperative actions in a BG with as a function of the number of generations for $g = 4$, and $F = 0.75$. The remaining parameters are the same of the paper by Suzuki and Akiyama (Suzuki and Akiyama 2005). Each curve averaged over 1000 realizations

is positive. In case the population contains no players with a good reputation, a candidate with a negative image score is accepted in the group. We also tested an alternative mechanism for partner selection: in a set of simulations we set the threshold for accepting candidates as being equal to or larger then the average strategy of the initial member of the group, but this did not produce any appreciable effects on the outcome of the simulations. Once the network is completely defined, each group plays a round of the game, with the same rules working on CG and SWN. The procedure (network construction followed by a round of the game of each group) is repeated ten times, then the evolution process takes place again following the same rule given of the previous cases.

In Fig. 6.8 and 6.9 we show the behavior of the model for $N = 200$ (or the closest integer compatible with the remaining parameters), $F = 0.75$, while keeping the other parameters equal to the ones used by Suzuki and Akiyama (Suzuki and Akiyama 2005).

Our results show that the final cooperation level is lower in the BG then in the CG case when the additional members of the groups are selected at random. However, when reputation-based partner selection is available in a population distributed on a bipartite graph, full cooperation is reached in a very short amount of time (about ten generations), and this is true also for large groups with 20 individuals ($g = 20$ in figure). This result does not depend on the way in which groups are assorted: even when partner selection is restricted to a small percentage of agents, it can favour

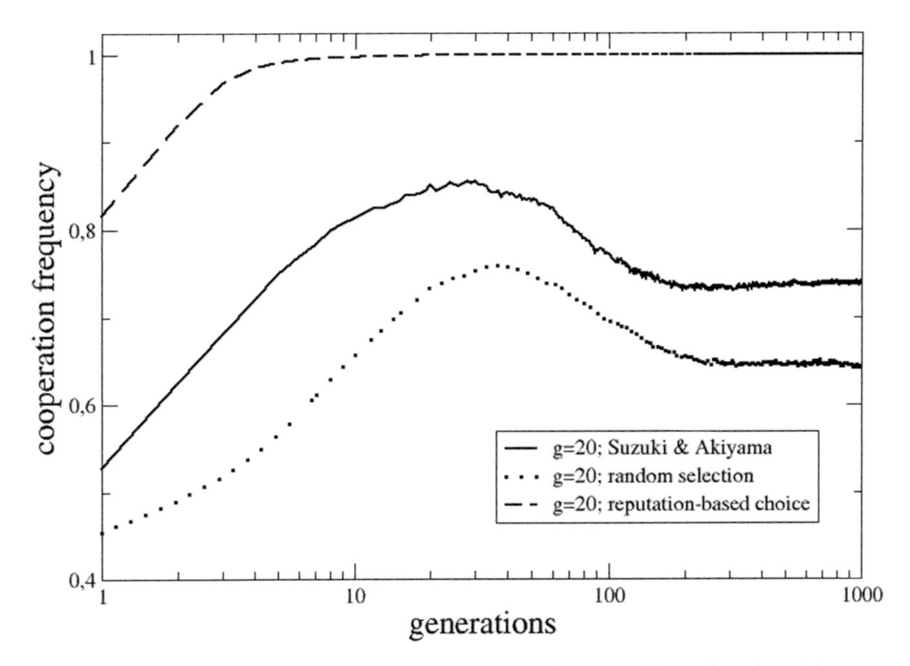

Fig. 6.9 Behaviour of the frequency of cooperative actions in a BG with as a function of the number of generations for $g = 20$ and $F = 0.75$. The remaining parameters are the same of the paper by Suzuki and Akiyama (Suzuki and Akiyama 2005). Each curve averaged over 1000 realizations

the invasion of the cooperative strategies throughout the system. This effect can be explained by the fact that, in general, in PGG it is better for individuals to get involved in as many groups as possible in order to maximize their income (Hauert and Szabo 2003). However, if this is not linked to a reputation-based partner selection mechanism, defection is still very profitable and cooperators are driven out of the system. On the contrary, if reputation is used to select group members, having a positive image score has a positive effect on fitness.

In the model by Nowak and Sigmund (Nowak and Sigmund 1998b; Nowak and Sigmund 1998a), based on the same image score mechanism, when the system ends up in a final configuration of complete cooperation, the only surviving strategy is usually $k=0$, that is, the "winning" strategy is a rather moderately generous one. A similar behavior appears with our model in CG and SW topologies.

On the other hand, when working on BG topology, the final system configuration, always totally cooperative, presents all the negative strategies, i.e. the more cooperative ones, as shown in Fig. 6.10. This means that taking into account realistic properties of social interactions among individuals not only makes cooperation spread throughout the whole population, but it also allows the survival of the most generous and altruistic strategies.

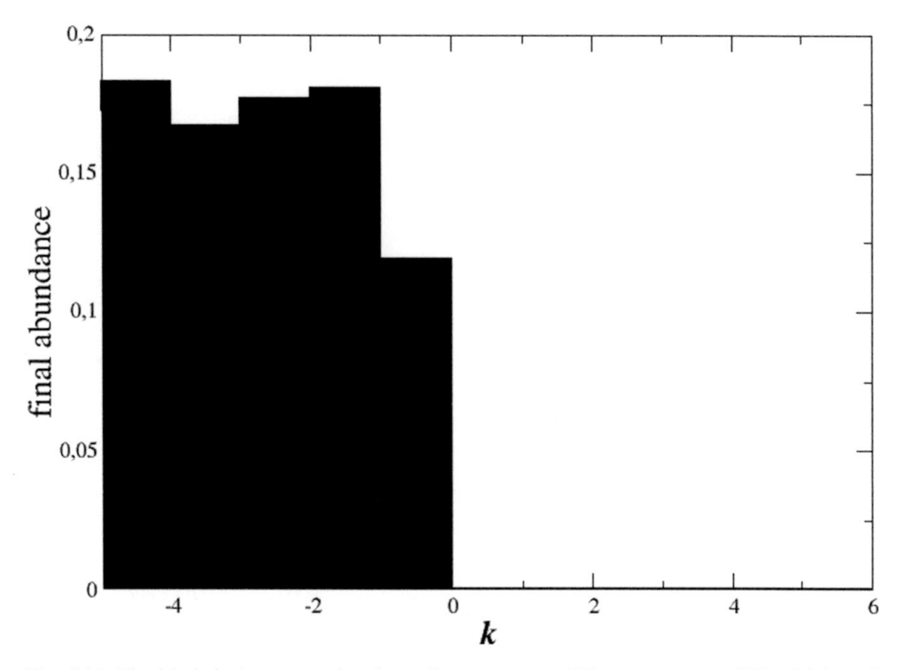

Fig. 6.10 Final (relative) strategy abundance for a system on BG, same system of Fig. 6.8 ($g = 4$) with reputation-based choice of the added members of each group. Values averaged over 1000 realizations

6.4 Discussion

In a PGG in which the history of agents' past interactions is publicly available as an image score, cooperation can emerge and be maintained for small groups of agents. When we move from a mean-field situation to a small-world network, we observe that cooperation becomes stable after 100 generations and for groups of four individuals. The real improvement is achieved thanks to the introduction of a partner choice mechanism on a bi-partite graph, where if a small percentage of group members are chosen on the basis of their reputations, cooperation can thrive.

In a social dilemma the introduction of a reputation mechanism for partner selection on a bipartite graph makes deception unprofitable, thus cooperators can thrive. In such an environment, agents with a positive reputation are more socially desirable, thus they can enter several groups in which their contributions help to achieve the social optimum. On the other hand, defectors with negative reputations are actively avoided, thus driving them to complete extinction after ten generations. Even more striking is the fact that, unlike other models (Suzuki and Akiyama 2005; Brandt et al. 2003), full cooperation is maintained even when group size increases.

6.5 Conclusions

The puzzle of the evolution of cooperation in humans can be successfully addressed if we take into account features of human societies that could have paved the way for the emergence of cooperative behaviors, like social networks and reputation. Moving from a replication of Suzuky and Akiyama (Suzuki and Akiyama 2005), we showed that cooperation can emerge and be maintained in groups of agents playing a PGG on a network. We used two network topologies with different group and total population size, finding interesting differences especially in terms of the maximum level of cooperation achieved. Our results show that when partner selection is available in an affiliative network, cooperation can be easily reached even in large groups and for large system's size.

The importance of social institutions (Ostrom 1990) and informal social control (Giardini and Conte 2012; Besnier 2009) is well known to social scientists who, like Ellickson (Ellickson 1991), have stressed the importance of these features: "A close-knit group has been defined as a social network whose members have credible and reciprocal prospects for the application of power against one another and a good supply of information on past and present internal events [...]. The hypothesis predicts that departures from conditions of reciprocal power, ready sanctioning opportunities, and adequate information are likely to impair the emergence of welfare-maximizing norms" (p. 181).

Introducing a small world network does not alter the dynamics of cooperation in a PGG in a fundamental way, and this is also true for a bipartite graph with random partner selection. However, when we model the world as made of groups that can actively select at least one of their members, cooperators outperform free-riders in an easy and fast way. The evolutionary dynamics of our model can be linked to a proximate explanation in psychological mechanisms for ostracism and social exclusion, two dreadful outcomes for human beings (Abrams et al. 2005; Baumeister and Leary 1995). In large groups of unrelated individuals, direct observation is not possible, and usually records of an individual's past behavior are not freely and publicly available. What is abundant and costless is gossip, i.e., reported evaluations about others' past actions, that can be used to avoid free-riders, either by refusing to interact with them, or joining another crew in which free-riders are supposedly absent. For this reason we plan to run simulations in which agents will be able to report private information about their past experiences, thus overcoming the unrealistic limitations posed by image score. We posit that the combination of a bi-partite graph social structure and gossip like exchanges will mimic human societies better and will provide useful insights about the evolution of cooperation in humans.

Acknowledgments We gratefully acknowledge support from PRISMA project, within the Italian National Program for Research and Innovation (Programma Operativo Nazionale Ricerca e Competitività 2007–2013. Settore: Smart Cities and Communities and Social Innovation Asse e Obiettivo: Asse II—Azioni integrate per lo sviluppo sostenibile).

References

Abrams, D., Hogg, M. A., & Marques, J. M. (2005). *The social psychology of inclusion and exclusion*. New York: Psychology Press.

Alexander, R. (1987). *The biology of moral systems (foundations of human behavior)*. New York: Aldine Transaction.

Baumeister, R. F., & Leary, M. R. (1995). The need to belong: Desire for interpersonal attachments as a fundamental human motivation. *Psychological Bulletin, 117,* 497–529.

Besnier, N. (2009). *Gossip and the everyday production of politics*. Hawaii University Press.

Brandt, H., Hauert, C., & Sigmund, K. (2003). Punishment and reputation in spatial public goods games. *Proceedings of the Royal Society of London. Series B, 270,* 1099–1104.

Carpenter, J. P. (2007). Punishing free-riders: How group size affects mutual monitoring and the provision of public goods. *Games and Economic Behavior, 60,* 31–51.

Diestel, R. (1997). *Graph theory*. Springer (4th Ed 2010).

Ellickson, R. (1991). *Order without law*. Harvard University Press.

Fehr, E., & Gachter, S. (2000). Cooperation and punishment in public goods experiments. *American Economic Review, 90,* 980–994.

Giardini, F., & Conte, R. (2012). Gossip for social control in natural and artificial societies. *Simulation, 88,* 18–32.

Gómez-Gardeñes, J., Romance, M., Criado, R., Vilone, D., & Sánchez, A. (2011). Evolutionary games defined at the network mesoscale: The public goods game. *Chaos, 21,* 016113.

Hardin, G. (1968). The tragedy of the commons. *Science, 162,* 1243–1248.

Hauert, C., & Szabo, G. (2003). Prisoner's dilemma and public goods games in different geometries: Compulsory versus voluntary interactions. *Complexity, 8,* 31–38.

Kiyonari, T., & Barclay, P. (2008). Cooperation in social dilemmas: Free-riding may be thwarted by second-order rewards rather than punishment. *Journal of Personality and Social Psychology, 95,* 826–842.

Maier-Rigaud, F. P., Martinsson, P., & Staffiero, G. (2005). Ostracism and the provision of a public good, experimental evidence, working paper series of the Max Planck institute for research on collective goods, vol. 24, Max Planck Institute for Research on Collective Goods.

Milinski, M., Semmann, D., & Krambeck, H. J. (2002). Reputation helps solve the tragedy of the commons. *Nature, 415,* 424–426.

Nowak, M. A., & Sigmund, K. (1998a). The dynamics of indirect reciprocity. *Journal of theoretical biology, 194,* 561–574.

Nowak, M. A., & Sigmund, K. (1998b). Evolution of indirect reciprocity by image scoring. *Nature, 393,* 573–577.

Ostrom, E. (1990). *Governing the commons: The evolution of institutions for collective action*. Cambridge University Press.

Ostrom, E. (2005). *Understanding institutional diversity*. Princeton University Press.

Panchanathan, K., & Boyd, R. (2004). Indirect reciprocity can stabilize cooperation without the second-order free rider problem. *Nature, 432,* 499–502.

Smith, E. A. (2010). Communication and collective action: The role of lan-guage in human cooperation. *Evolution and Human Behavior, 31,* 31–245.

Suzuki, S., & Akiyama, E. (2005). Reputation in the evolution of cooperation in sizable groups. *Proceeding of the Royal Society B, 272,* 1373–1377.

Walker, J. M., & Williams, A. (1994). Group size and the voluntary provision of public goods: Experimental evidence utilizing large groups. *Journal of Public Economics, 54*(1), 1–36.

Watts, D. J., & Strogatz, S. H. (1998). Collective dynamics of small-world networks. *Nature, 393,* 440–442.

Wedekind, C., & Milinski, M. (2000). Cooperation through image scoring in humans. *Science, 288,* 850–852.

Part II
New Applications

Chapter 7
A Novel Interdisciplinary Approach to Socio-Technical Complexity

Sociologically Driven, Computable Methods for Sport-Spectator Crowds' Semi-Supervised Analysis

Chiara Bassetti

7.1 Introduction

Contemporary societies are more and more characterized by the presence of large, complex, and sometimes overlapping socio-technical systems (e.g., Button and Sharrock 1998; Suchman 2002; Avgerou et al. 2004; Whitworth 2009)—that is, systems embedding human and artificial agents, and incorporating as a whole technology and practice, bodies and tools, place and activity. "The complexity of such systems makes it very hard for them to cope with critical situations, as they emerge from the interplay between all participants and cannot be reduced to mere technical malfunctioning or to the negligence/malevolence of the human actors involved" (Ferrario 2011).

To deal with such a complexity, we need, first, an interdisciplinary approach able to tackle the issue at the systemic level without losing its empirical ground and practiced-based, micro-analytical perspective. Second, with special reference to critical situations, one promising avenue to follow is designing artificial agents able to take into consideration—i.e., to detect and reason about—not only physical/sensorial, nor only cognitive, but also social aspects of a scene. Ultimately, we want technologies that are able to participate in social interaction and everyday human activities much like humans do. Indeed, if one considers critical situations, this could even make a difference in avoiding catastrophes (cf. Bassetti et al. 2013).

To move toward such objectives, we propose an interdisciplinary approach which—by leveraging on sociology, especially micro-sociology and ethnomethodology (EM/CA approach—see Sect. 7.3), computer vision, particularly Social Signal Processing (SSP) (e.g., Vinciarelli et al. 2009; Cristani et al. 2010; Setti et al. 2015), and foundational and applied ontology (e.g., Guarino 1998, 2009)—aims at the integration of microsociological analysis into computer-vision and pattern-recognition

C. Bassetti (✉)
Laboratory for Applied Ontology, Institute of Cognitive Sciences and Technologies – CNR, Via alla Cascata 56 C, 38123 Trento, Italy
e-mail: chiara.bassetti@loa.istc.cnr.it

F. Cecconi (ed.), *New Frontiers in the Study of Social Phenomena*,
DOI 10.1007/978-3-319-23938-5_7

modeling and algorithms, both "directly" and via the mediation of ontology. In addressing the need for interdisciplinary methods to deal with socio-technical complexity at a systemic yet micro-grounded level, the approach, which this chapter shall present in detail as applied to the analysis of a particular setting, and especially as for its sociological foundations, is—

- both theoretically and analytically driven;
- empirically grounded and concerned with the minute details of naturally occurring social interaction;
- yet systemic and multidimensional (in considering individuals, groups and subgroups; human and artificial agents);
- semi-supervised and computable;
- and oriented toward large-scale applications.

The general idea behind the proposed approach lays at the basis of the project VisCoSo[1] that we are conducting at the Laboratory for Applied Ontology (LOA-ISTC-CNR), and that has an international airport as its core case study to analyze critical situations in socio-technical systems. However, we have applied the same approach to another project, OZ (Osservare l'attenZione), which focuses empirically on what is commonly referred to as sport-spectator crowds. This second setting, on which the chapter shall focus, is as (i) complex, (ii) socio-technical, and (iii) prone to critical situations and emergencies as the former, yet it constitutes a simpler case to test our approach given some of its material and organizational characteristics—*in primis*, people's location being mostly constrained to the stands.

Crowds, better defined as "large gatherings" (Goffman 1961, 1963; McPhail 1991) are almost ever-present in our societies: from urban spaces to airports, from ERs to malls, from churches to theaters and arenas, we are immersed in crowds of different kinds every day (see Sect. 7.3.2). Therefore, capturing and understanding crowd dynamics is crucial, and this is true under diverse perspectives. From social sciences (sociology, social movement studies, political science, organization studies, etc.) to public safety management and emergency response practices, modeling and predicting large gatherings' presence and dynamics, thus possibly preventing critical situations and/or being able to properly react to them, is fundamental. This is where semi/automated technologies can make the difference. The work presented in this chapter, focused on spectator crowds, is intended as a scientific step toward such an objective.

In what follows, I shall introduce the OZ project and the multi-step method we employed (Sect. 7.2) and then focus on some of such steps—namely, those that are more concerned with the sociological contribution. Section 7.3 also presents the theoretical basis of the proposed approach from a sociological point of view; as for regarding the state of the art in human and crowd behavior analysis from a computer vision perspective, I make reference to previous work (Conigliaro et al. 2015). Section 7.4 focuses on the EM/CA video analysis of the empirical material that I conducted as a basis for the selection of what I call the *atomic components of*

[1] http://www.loa.istc.cnr.it/projects/viscoso.

action-in-interaction in the considered setting (i.e., in sport-spectator gatherings). Such components, together with the formal compositional methods elaborated for having them used in computable, automated tasks, are presented in Sect. 7.5. Finally, some reflections are made concerning the scientific as well as applicative advantages of the approach, and the avenues it opens to meet contemporary societal challenges (Sect. 7.6).

7.2 The OZ Project at a Glance

The interdisciplinary project OZ—*Osservare l'attenZione* (*Observing attention*)—was conducted in the course of the 2013 Trentino Winter Universiade by the LOA-ISTC-CNR, the Laboratory of Vision, Image Processing & Sound (VIPS) of the University of Verona, and the Department of Information Engineering and Computer Science (DISI) of the University of Trento. The aim was to develop a technology able to automatically detect at run time spectators' attention and excitement levels. More specifically, the detailed objectives were the following:

1. Spectators segmentation: distinguishing fan groups belonging to different teams (or just attending), even when they are merged in the stands; finding diverse groups among spectators (e.g., attentive vs. distracted, enthusiastic vs. bored spectators).
2. Attention and excitement level calculation: in a given time interval, quantizing the level of attention and excitement of the crowd or some of its parts.
3. Event segmentation: single out the most salient events of the match, like goals, fouls, or shots on goal on the basis of diverse crowd activities (e.g., clapping hands when the favorite team scores a goal, getting excited when a foul is or is not signaled by the referee).

As I mentioned, the chosen path to reach such objectives has been the integration of it should be micro-sociological analysis into computer-vision and pattern-recognition modeling and techniques. The process (Fig. 7.1) was the following: While preparing the theoretical background (sociology, social psychology, social ontology), we built a novel repository of videos taken during the 2013 IIHF Ice Hockey U18 World Championship (Asiago, April 7–13, 2013). The videos, on the one hand, have been used as empirical material for the sociological analysis on whose basis the atomic components of action-in-interaction where selected (notation). On the other hand, they served to test the framework from a computer-vision perspective. The successful results of the study (cf. Conigliaro et al. 2013a, b) allowed us to apply the approach, by then enhanced through sociologically driven annotation, to the four final hockey matches of the 2013 Winter Universiade, and to release the related, annotated S-HOCK (Spectator-HOCKey) dataset (cf. Conigliaro et al. 2015). The enhanced approach was found to outperform the previous one. Simultaneously, leveraging on both sociology and ontology, we elaborated some formal methods of composition of the above-mentioned atomic components, to be integrated into

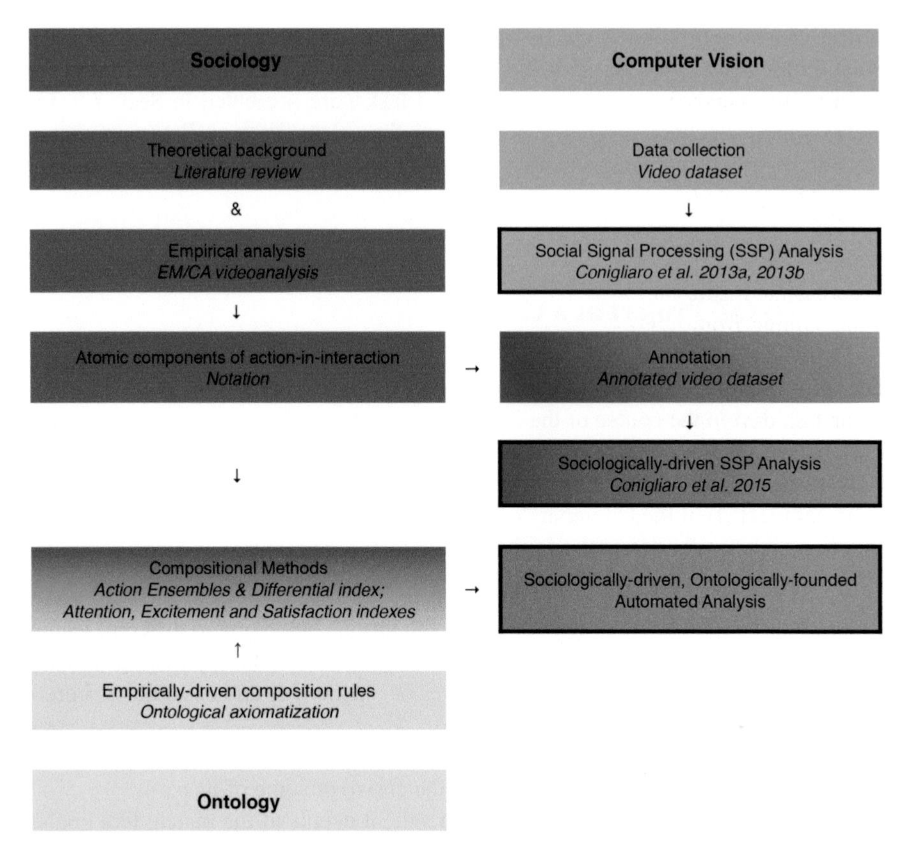

Fig. 7.1 Scheme of the interdisciplinary approach taken to analyze sport spectator crowds

computer-vision and pattern-recognition algorithms. We are now in the process of testing the performance of this third approach, in addition to enlarging the dataset with some basket matches.

7.3 State of the Art and Theoretical Foundations

Among the whole set of possible features that can be extracted from a video recording, we have selected the annotated "elementary forms of action" (McPhail 1991; Schweingruber and McPhail 1999)—or atomic components of action-in-interaction—as strictly connected with the analysis of social interaction, and related to our specific setting, i.e., sport-spectator crowd. Whereas the relevance of body posture and motion in interpersonal interaction is well known (e.g., Birdwhistell 1970) and the same holds for proxemic formations (Kendon 1990) and participation frameworks (Goffman 1961, 1963, 1967), other actions, such as jumping, waving arms, or shaking "fan objects," are specific to certain settings. In both cases, behavior must be considered with attention to its situated context.

From this perspective, we drew from available literature on—

1. social interaction, with particular attention to non-verbal conduct (proxemics, body posture, gesture, etc.), especially in public places (Goffman 1963);
2. so-called (but cf. McPhail 1991) "crowd behavior"—i.e., social interaction in large gatherings, in particular sport-spectator gatherings.

Whereas the former body of literature is huge and diverse, especially for contributions coming from micro-sociology (e.g., Goffman 1961, 1963, 1967) and Ethnomethodology and Conversation Analysis (EM/CA) (e.g., Garfinkel 1967; Kendon 1990, 2004; Sacks 1992), the latter, with particular regard to the sportive setting, is far less developed and often connected to surpassed theories on/of "crowd behavior" and "collective behavior" (cf. McPhail 1991; cf. also Blumer 1951; Goode 1992). I shall now focus on these two bodies of literature.

7.3.1 (Multimodal) Social Interaction

While the term was only introduced some years ago, micro-sociology's roots can be found in "classical" scholarly work, starting from Simmel's (1908) attention to the smallness and elusiveness of some forms of sociability (from flirtation to gossiping, from silence to the role of the senses in everyday interaction, etc.) coupled with their being ubiquitous and practiced by everybody—the so-called micro-macro link. Simmel's focus on the inconspicuous mechanisms through which society is engendered has then been taken up by the ethnographers of the so-called "Chicago school," from which two main approaches devoted to the study of collective behavior and interaction have emerged: (i) the "ecological" one of urban sociology (e.g., Park et al. 1925), especially concerned with subcultures and deviance; and (ii) symbolic interactionism (e.g., Mead 1934; Blumer 1969), whose basic premise is that people act toward things and respond to actions of others based on the meaning those things or actions have for them, and those meanings are symbolically mediated and are realized in social interaction. Although it is highly debatable whether Goffman (e.g., 1959, 1961, 1963) can be properly included into symbolic interactionism (e.g., Collins 1988), he is certainly one of the most renowned exponent of the interactionist approach in recent sociology—actually, he is the one who succeeded in making everyday rituals and face-to-face interaction legitimate phenomena of sociological interest, and the one who elaborated the terminology and definitions that made such phenomena "sayable." McPhail's work on large gatherings, while standing at some distance from Blumer's conception of collective behavior, heavily rests on Goffman's framework for the analysis of social interaction.

Such a framework's theoretical roots are to be found, not in symbolic interactionism, but another "classical" scholar's thought: Durkheim's (for example, 1965). And it is not by chance that the Durkheimian tradition lays at the foundations of

Garfinkel's (1967, 2002, 2006) ethnomethodology (EM) (cf. also Rawls 2011). The latter's theory of social action is not only as much concerned with (local) morality as both Durkheim's and Goffman's theories are, but it is also "radical" in pursuing the study of the orderliness of ordinary phenomena as an everyday collective achievement—that is, as achieved every day during and through situated interaction. The analysis of the orderliness of conversation, or talk-in-interaction, has been then the specific challenge taken up by conversation analysis (CA) (e.g., Sacks 1992; Psathas 1995; Heritage 1999). Since the seventies, CA has not stopped studying the ordered features of talk—both institutional and ordinary/informal, both face-to-face and mediated—and then, thanks also to the opportunities opened up by video recordings, of gestures, proxemics, and nonverbal behavior in general, the full (video)analysis of multimodal interaction (e.g., Mondada 2008; Heath et al. 2010) and its application to manifold empirical contexts.

The approach, which now goes bythe EM/CA label, can leverage on a large and variegated research corpus on nonverbal conduct that has been particularly helpful for the OZ project. Birdwhistell's (1952, 1970) work on human movement—or *kinesics*, as he coined it—goes back to the fifties and has then been followed by that of various other scholars (e.g., Kendon 1970, 1972; Knapp 1972; Duncan and Fiske 1977; Dittman 1987; Burgoon et al. 1989). The study of body language and nonverbal communication has been ongoing since the sixties (e.g., Scheflen 1964, 1972; Bateson 1968; Ekman and Friesen 1969b), and particular attention has been payed to posture (e.g., Scheflen 1964; Mehrabian 1969; Kendon 1970; LaFrance and Broadbent 1976; Matsumoto and Kudoh 1987), proxemics and "formations" (e.g., Hall 1963, 1966; Ciolek and Kendon 1980; Kendon 1992; McPhail 1994)[2], and self-touching, or self-manipulation (e.g., Ekman and Friesen 1969a; Rosenfeld 1973)—that is, in gestures studies' jargon, "adaptors." A whole, largely interdisciplinary scientific area, indeed, is devoted to the analysis of gestures and their role in interpersonal multimodal interaction (e.g., Sapir 1927; Ekman and Friesen 1972; Lefebvre 1975; Kendon 1990, 2004; McNeill 2005)—an endeavor in which computer vision participates too (cf. Rautaray and Agrawal 2015). Scholarly attention, finally, has been also caught by head and body orientation as well as gaze and the role it plays in sociability (e.g., Nielsen 1962; Argyle and Dean 1965; Kendon 1967; Knapp 1978; Hietanen 2002)—an interest, the latter, that traces as back as to Simmel.

To conclude, some more detailed issues, relevant for the research at hand, are listed below.

- There is some evidence that the *closed fist* constitutes a widespread gesture of power and triumph (Morris 1994, p. 70).
- An early experimental study by James (1932), based on ratings by judges, identified four *postural categories*: (a) forward lean ("attentiveness"); (b) drawing back or turning away ("negative," "refusing"); (c) expansion ("proud," "conceited," "arrogant"); and (d) forward-leaning trunk, bowed head, drooping shoulders, and sunken chest ("depressed," "downcast," "dejected") (cf. Mehrabian 1972, p. 19).

[2] We tried the integration of micro-sociology and computer vision also with respect to the analysis and detection of "free-standing conversational groups" and "facing formation" (Setti et al., 2015).

- Research on *sitting positions* has been carried out by Hewes (1957) as well as Scheflen (1972). Vrugt and Kerkstra (1984) found female, North American college students showing uneasiness by sitting still and arm-crossing.[3] According to Morris (1994, pp. 152–54), there are four typical human sitting postures in chairs: (a) ankle–ankle legs cross ("I am politely relaxed," worldwide), (b) knee–knee legs cross ("I am very relaxed," worldwide), (c) ankle–knee legs cross ("I am assertively relaxed," widespread), and (d) legs twined ("I am slinkily relaxed," widespread).
- Akimbo (or *hands-on-hips*) position, in which the palms rest on the hips with the elbows flexed outward, was identified as a human "posture type" by the anthropologist Hewes (1957). Mehrabian (1969) later found that in standing hand-on-hips interactants, akimbo was used more with disliked than with liked partners. Moreover, "arms-akimbo position is more likely when you are talking to a person you see as having a lower status than your own" (Knapp 1972, p. 101). According to Morris (1994, p. 4), arms akimbo "is an unconscious action we perform when we feel anti-social in a social setting. It is observed when sportsmen have just lost a vital point, game or contest." Finally, one- and two-handed, stylized versions of the akimbo posture are used by African American girls and women to show anger, disgust, and disagreement (Givens 2002).
- It appears that in conditions of severe crowding, the frequency of *arms crossed* in front of the body touching at the crotch greatly increases (Baxter and Rozelle 1975, p. 48)

7.3.2 Large Gatherings (or Crowds)

The main reference for crowd studies is McPhail's extensive work (e.g., McPhail 1991; Schweingruber and McPhail 1999) on *large gatherings* (ranging from demonstrations to sport events) and his taxonomy of "elementary forms of action", which we adapted to our specific needs—i.e., those concerning the context of activities (and the results of preliminary EM/CA video analysis), those related to the video corpus (e.g., facial expression not detectable), and that of a good compromise between taxonomical accuracy and activity cost in annotating.

Let us start by considering McPhail's (1991, p. 159) definition of collective action:

> Two or more persons engaged in one or more actions (e.g., locomotion, orientation, vocalization, verbalization, gesticulation, and/or manipulation), judged common or concerted on one or more dimension (e.g., direction, velocity, tempo, or substantive content).

There are three main issues to point out. First, within the first set of parentheses are the six main categories of McPhail's taxonomy. We have taken them all into consideration except for vocalization and verbalization, as we were not analyzing the audio of the recordings in computer-vision terms. For the sake of the annotation manual's efficacy, the remaining four categories of action and their member-items

[3] On gender and bodily conduct see also, e.g., Young (1980) and Guillaumin (1992).

selected for annotation have been regrouped according to more computer-vision-like criteria (e.g., orientation as "head pose," gesticulation and manipulation in the same large category as "action").

Second, the first half of the quote (until "judged…") is a good general definition of social interaction (or action-in-interaction, in EM/CA terms) that may or may not involve common or concerted—i.e., collective—action. In Goffmanian (1963, 1981) terms, we should talk of unfocused interaction (i.e., individual action in public), common-focused interaction (e.g., watching a movie or a match together), and jointly focused interaction (e.g., cheering, conversing).

Third, it is important to bear in mind that social interaction is a multi-scalar entity: individual, common, and concerted actions-in-interaction are often co-occurrent and mutually intertwined in everyday life situations (e.g., conversing with a partner while attending a theater show). Audiences, or spectator crowds, point to a particular kind of interaction—i.e., common-focused—and a particular kind of collective action—i.e., common— yet they often also involve unfocused as well as jointly focused interaction. This has been taken into consideration, for instance, by annotating people walking (away from the stands) or pointing toward something outside the game field to the benefit of a nearby companion, or kissing each other.

It is now time to introduce a taxonomy of crowds—or large gatherings—to better identify the one we are dealing with:

- *prosaic* (McPhail 1991) or *casual* (Blumer 1951; Goode 1992) crowds, where members have little in common except their spatio-temporal location (e.g., line at the airport check-in counter);
- *demonstration/protest* (McPhail 1991) or *acting* (Blumer 1951; Goode 1992) crowds, a collection of people who gather for specific protest events (e.g., mob/riot/sit-in/march participants);
- *spectator* (McPhail 1991; cf. also Berlonghi 1995) or *conventional* (Blumer 1951; Goode 1992) crowds, a collection of people who gather for specific social events (e.g., cinema/theater/sport spectators);
- *expressive* (Blumer 1951; Goode 1992) crowds, a collection of people who gather for specific social or ritual events *and* want to be full members of the crowd, to participate in "crowd action" (e.g., flash-mob dancers, Mass participants, sport supporters).

It is important to notice that different types of crowd can be co-present (e.g., a flash-mob at the airport while others are in line) and even intertwined, which is precisely the case of sport supporters within a broader sport audience. For the purposes of our research, therefore, we consider expressive crowds as sub-parts of spectator crowds. It may be worth mentioning that all computer-vision approaches assume a general and unique kind of crowd, and focus primarily on casual and protest crowds (cf. Conigliaro et al. 2015).

Moving from crowds in general to *spectator crowds* in particular, one should consider some scholarly work on one-to-many speaking situations, such as Goffman's (1981) work on conference lectures, and some EM/CA analyses of political meetings and confrontations, and of the close relationship between applause and *performance*—more than the speech content or the speaker's popularity/status (e.g., Atkinson 1984; Heritage and Greatbatch 1986). More generally, a part of the anthropological literature focused on the performer–audience relation in theaters and during rituals (e.g., Schechner 1971, 1986). Other research focused on one-to-many speaking situations, yet from an experimental rather than empirical perspective, studied intra-audience effects (e.g., Hylton 1971; Hocking et al. 1977; Hocking, 1982). Hylton (1971) showed that naïve audience members exposed to positive audience response were more favorable toward the speech topic and the speaker than those exposed to negative audience feedback; similarly, Hocking's (1982) research in rock-and-roll bar situations showed that those exposed to positive response evaluated the band more positively, stayed longer at the bar, and had a greater desire to see the band again.

Turning to the specific domain of *sport-spectator crowds*, we considered sociological and psychological literature, and we found four main research areas:

- the risk the context entails for violence (e.g., Goldstein and Arms 1971; Mann 1979; Roadburg 1980)[4];
- spectators involvement, motivation, and satisfaction (e.g., Sloan 1979; Zillman et al. 1989; Yiannakis et al. 1993; Madrigal 1995; Kerstetter and Kovich 1997; Choi et al. 2009; McDonald et al. 2002; Bowker et al. 2009);
- intra-audience effects, particularly concerning excitement/enjoyment, attendance, and event evaluation (e.g., Hocking 1982);
- audience-team effects, such as the "home advantage" (e.g., Schwartz and Barsky 1977; Allison 1979; Edwards 1979; Madrigal and James 1999).

We neglected most of this literature—especially that for the first two areas—since it deals either with surpassed theories of collective behavior (e.g., "contagion" theory—cf. Hocking 1982; Levy 1989) or with motivational factors that—even when properly measured and accounted for—are good predictors of long-term socio-psychological involvement more than situated behavioral involvement (cf. also Choi et al. 2009); and yet the latter is the only one accessible by the visual means of computer vision and by EM/CA methods alike, since the discipline's epistemological foundations (cf. Garfinkel 1967). However, some studies are worthy of notice.

Facing the issue of intra-audience effects, after having considered "contagion", "convergence", "emergent norm", and "informational" theories, Hocking (1982) concluded that there was still a need for "research providing empirical support for the thesis that audience response to sporting events affects other audience members' arousal/excitement/enjoyment, attendance, and evaluations of the event itself" and that such a research "would need to take into account a range of variables." Variables are listed in Table 7.1—some of them operate alone, others interact (cf. Mann 1979).

[4] "A persistent and popular view holds that high population density inevitably leads to violence. This myth, which is based on rat research, applies neither to us nor to other primates" (de Waal et al. 2000, p. 77).

Table 7.1 Intra-audience effects: Relevant variables (elaborated from Hocking 1982)

Variable relevant w.r.t. intra-audience effects	Greater effect state of the variable
Crowd size	Large crowd
Crowd density	Tightly packed crowd
Response intensity/volume	High intensity/volume
Indoor vs. outdoor arena/stadium	Indoor arena/stadium
Arena/stadium design	Field-stands proximity
Standing vs. sitting audience	Standing audience
Audience mutual coordination	High and continuous mutual coordination
De/inviduation	Anonymity

As for audience-team effects, Grusky (1963, p. 60) correctly pointed out that "enthusiastic support from the crowd may stimulate the player to put out more in the same way that a responsive audience can help produce scintillating dramatic performances on stage." In this regard, it is worth mentioning a study by Schwartz and Barsky (1977), who conducted comparative empirical research on football, baseball, hockey, and basketball and found that the home advantage is more pronounced—

- where arena/field variations are least conspicuous and conditions of play are therefore most uniform—i.e., where variables other-than-support are less relevant, and this means that the home advantage is a function of *social support* primarily;
- for indoor sports (hockey especially)—i.e., where the setting itself tends to make the support more *immediate and intense* thanks to the audience's compactness and proximity to the game field;
- for sports characterized by continuous rather than intermittent game-action and consequent supportive-action (hockey and basketball vs. football and baseball)—i.e., where there is *sustained* support.

It is not by chance that the home team presents a greater level of offensive (rather than defensive) activity, and the advantage is largely traceable to superior offensive performances, "[f]or these are precisely the kinds of activities most likely to elicit the approval" (1977, p. 652), and for—as the video analysis revealed (see Sect. 7.4)—these are much more recognizable as salient.

7.4 Empirical Foundations

Besides the above-illustrated literature basis, the annotation's items selection has been driven by the analysis of the video-set performed accordingly to the principles and procedures of EM/CA video analysis (e.g., Heath et al. 2010). Preliminary analysis has identified two main activities enacted by sport spectators:

1. *reading the field*, that is, game-actions' projection;
2. *performing the stands*, which can be further divided into
   a. doing [attending the game],
   b. doing [supporting the team].

Table 7.2 Spectators' activities and related groups of markers

Activities		Markers of
Reading the field		Game-action projection
Performing the stands	Doing [attending the game]	Dis-attention/engagement with game-field activities
		Mutual coordination
	Doing [supporting the team]	Enjoyment/annoyance and dis/satisfaction with game-action and game-action's outcome, respectively
		Mutual coordination

Table 7.3 Dis/attention and dis/engagement markers

Attention/engagement markers	Dis-attention/engagement markers
Head/gaze toward the field	Head/gaze toward a fellow spectator or downward (e.g., to one's phone, camera, purse)
High chin [less significative]	Low chin [less significative]
Hands (open palm) or elbows on knees[a]	With one's arms folded or idle hands
Torso inclined toward the field (less than 90° angle between torso and legs)[a]	Reclined chest (on the back of the seat if present)[a]
Upright torso, straight shoulders and absence of abdominal contraction	Not fully upright chest, curved shoulders and/or abdominal contraction
Both feet on the ground[a]	Crossed legs[a]
Pointing toward something on the field for the benefit of a fellow spectator	Pointing toward something outside the field for the benefit of a fellow spectator
Moving body weight from one to the other gluteus[a]	

[a] When seated

Supporting the team in common-sense terms coincides with what I call *performing the stands*. Here, doing [supporting the team] has a stricter definition: it means *displaying* support (i.e., standing, jumping, clapping, etc.), just like doing [attending the game] means displaying attendance/attention (i.e., pointing to or looking at the game field).

With this in mind (see also Table 7.2), video analysis has been then devoted to the identification of markers of—

1. dis/attention and dis/engagement with the game-field activities (Table 7.3);
2. game-actions projection, with consequent increase in attention/engagement—that is, excitement (Table 7.4);
3. enjoyment/annoyance and dis/satisfaction with respect to, respectively, game-actions and game-actions' outcome (Table 7.5);
4. mutual coordination in doing both [attending the game] and [supporting the team] (Table 7.6), the latter equaling mutual coordination in displaying enjoyment/annoyance and dis/satisfaction with particular game-actions or game-actions' outcomes[5].

[5] Verbal conduct (e.g., chorus) is highly relevant in facilitating and enhancing intra-audience co-ordination and synchrony, yet it has not been considered here given the technical characteristics of the dataset.

Table 7.4 Markers of significant upcoming game-action's projection

Projection of significant upcoming game-action markers
Disclosing lips, opening mouth (with head oriented toward the field)
Taking a big breath in (inhalation)
Pelvis slightly rising from the seat, feet leaning on the ground
Moving from seated to standing position
Lifting camera up (to a point between diaphragm and eyes)
Lifting binoculars up and/or using them
Shifting from any status of dis-attention/engagement to the corresponding one of attention/ engagement (Table 7.3)

Table 7.5 Markers of enjoyment/satisfaction and annoyance/dissatisfaction with the game-action or its outcome

Satisfaction/enjoyment markers	Dissatisfaction/annoyance markers
Hopping (relevant: jumps' highness and number)	Hitting one hand against the other one once
Raising arm/s over one's head or opening arms (relevant: movement's largeness)	Hitting one hand with open palm on a thigh
Repeatedly moving arm/s (relevant: duration; secondarily, largeness and closed fist/s)	Bringing hand with open palm toward head (mouth, chick, forehead…) (relevant: posture duration)
Applauding (relevant: rhythm [fast/slow]; hands' position [high/low]; secondarily, movement's largeness [by wrist/ forearm] and duration)	Leaving arms falling on one's side or in one's womb (abandoning precedent posture and releasing muscular tension/energy)
Shaking fan-objects (relevant: movement's largeness and duration)	Lowering megaphone or camera
Putting hands in cone (or megaphone of some kind) in front of one's mouth	

Table 7.6 Markers of mutual coordination in doing [attending the game] and [supporting the team]

Mutual coordination in doing [attending the game]	Mutual coordination in doing [supporting the team]
Head & body orientation	Synchrony in applauding, clapping, etc.
Torso posture and, more generally, posture sharing	Homogeneity in shaking fan-objects: direction (e.g., everybody toward right then left), largeness, and rhythm
Sitting/standing position	Homogeneity in hopping: highness of the jump and rhythm
Pointing to something on the game field	Physical contact among spectators: hugs, pats on the back, etc.

From a sequential and processual point of view, it is important to notice that, with regard to the projection of significant upcoming game-actions, what follows is, alternatively—

a. performing one or more of the actions marking satisfaction/enjoyment or of those expressing dissatisfaction/annoyance, depending on the (outcome of the) considered game-action;
b. falling back to one or more of the statuses of attention/engagement (or, more rarely, disattention/disengagement) when, eventually, the projected game-action does not take place.

In Table 7.7, just as an example, the synthetic results of the sequential, second-by-second analysis of the video-recordings from an aggregated point of view. The first column contains the time reference; the second one, in bold, a synthetic description of (a segment of) the spectator crowd's "state," whereas the third one describes it in more detail; the fourth one in italics, finally, reports the co-occurrent ongoing situation on the game-field.

Reading the field and *performing the stands* are collective co-occurrent achievements. A couple of final considerations on such a matter could be drawn. First, *reading the field* allows *performing the stands*, or one will not be ready to produce the correct performance "on time". It is like waiting for producing one's line when acting, one's step when dancing: one should pay attention to what others are doing (and are going to do) in order to know *when* performing this or that action. However, unlike theatrical enactment and like instead ordinary conversation or artistic improvisation, there is no written plot. Therefore, one needs also to know *what* to perform at any given point, and action projection takes on more relevance. In brief, one is performing when attending to others' doings in order to know what to perform when. *Reading the field*, here, primarily means reading the game-field (direct attendance), but the stands-fields clearly offer cues as well (peripheral attendance). Second, *performing the stands* consists of two primary doings—i.e., [attending the game] and [supporting the team]—that are always co-present, only analytically distinguishable. However, they take different relative "weights" at different times (therefore, the spectator has to know what to primarily do when):

- doing [attending the game] is more relevant within one salient game-action and the following one, and especially *at the nascent state of salient game-actions*; it parallels attention (vs. disattention) markers and rises, like attention does, with action-projection;
- doing [supporting the team] becomes more relevant *during salient game-actions*, and "explodes" immediately *after their outcome*; it parallels dis/satisfaction and enjoyment/annoyance markers.

Figure 7.2 schematically represents the spectator-crowd—game-field dynamic. Notice that the scheme is representative of the whole spectator crowd (not just one segment).

Table 7.7 An example of sequential videoanalaysis for the largest segment of the crowd (Norway fans)

00.00.00	**Relaxed attention**	Almost all spectators are reading the field and doing [attending the game]	*Norway's offensive game-action starting from their half of the game field*
00.00.01	**Involved attention and salient game-action projection**	Increased coordination (e.g., heads orientation) and augmented display of doing [attending the game] (e.g., upright torso posture), while keeping reading the field	*Norway surpasses the half-field line*
00.00.05 00.00.06		Some spectators timidly do [supporting the team] with reference to the game-action's outcome (e.g., breathing out, torso moving slightly backward)	*Norway surpasses the 3/4 field line*
00.00.07	**Attention**	Slightly diminished coordination (e.g., heads orienting toward centre less synchronously); reading the field; doing [attending the game]	*Norway looses ball posses; Italy starts an offensive game-action, and reaches the central area of the game field*
00.00.12	**Involved attention and salient game-action projection**	Increased coordination; augmented display of [attending the game] (e.g., upright torso posture); keeping reading the field	*Norway takes ball possess back, and starts an offensive game-action*
00.00.14 00.00.15		Some spectators do a little [supporting the team] for the ongoing game-action (e.g., applauding), then go back to the "mere" status of involved attention with the other spectators, in a crescendo	*Norway surpasses the 3/4 field line*
00.00.18	**Crescendo in attention and salient game-action's outcome projection**	Increased attention in reading the field; increased doing [attending the game] and strong anticipation/preparation of/for doing [supporting the team] (e.g., upright torso posture, standing, fan objects got ready to be shaken)	*And keeps on with the offensive game-action*
00.00.19	**Displayed dissatisfaction for game-action's outcome**	Somebody do [supporting the team] but then come back with the others to the status-in-crescendo of involved attention (*interruptus* [supporting])	*Missed goal by Norway*
00.00.21	**Displayed satisfaction for game-action's outcome:**	Increased coordination; reading the field reaches almost zero; doing [supporting the team] towers over doing [attending the game]	*Goal by Norway* *"Pause" in game-action*
00.00.50	**Decreased doing [supporting** the team], until it completely stops		*Players starts moving again to the centre*
00.00.54 00.01.10	**Commenting phase**	Spectators exchange comments; they are looking in different directions (body and gaze orientation), since in this moment they are more concerned with and oriented to interpersonal interaction on the stands than with/to the game/field	
00.01.18	**Toward relaxed attention**		*Game-action restarts (with an offensive action by Italy)*

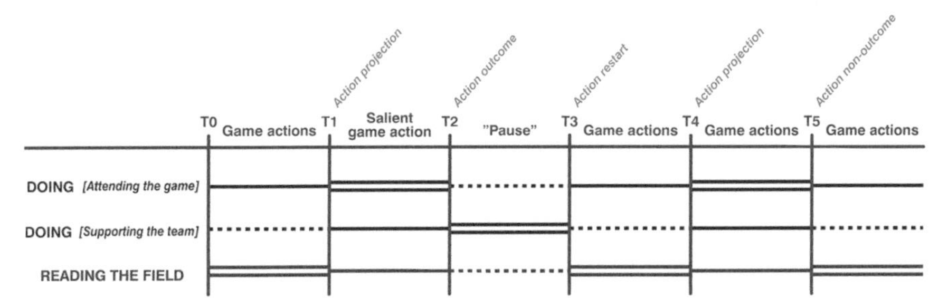

Fig. 7.2 The spectator-crowd—game-field dynamic. Different lines represent different "weights": *dashed line*: background activity; *single straight line*: middle-ground activity; *double straight line*: foreground activity

7.5 Atomic Components, Annotation, and Compositional Methods

Having completed both the theoretical and the empirical/analytical work, I selected a pool of atomic components of action-in-interaction relevant for the considered context (see Table 7.8[6]), on whose basis we proceeded to video annotation. Using the ViPER format (Doermann and Mihalcik 2000) and toolkit (http://viper-toolkit.sourceforge.net), each video sequence has been annotated frame by frame, spectator by spectator. Each annotator processed 930 frames, and was asked to do it in a lab. After all the sequences have been treated, producing a total amount of more than 100 millions of annotations, a second round started, with "second-round annotators" that had to find bugs in the first-round annotation. The whole work lasted almost 1 year and involved 15 annotators, all paid for their work.

As I mentioned, a first result of such a work has been the enhanced SSP analysis of the dataset, which was found to outperform the standard-method analysis (see Conigliaro et al. 2015). However, I proceeded in parallel to the elaboration of compositional methods—to be then axiomatized with the help of ontology so to make them computable—for the selected components. This basically amounted to the following activities.

1. With respect to the task of crowd segmentation (between-individuals analysis), I have created ensembles of "similar" components (Table 7.9), called *Action Ensembles*, serving as a basis for detecting, counting and grouping together those who are doing the same or a similar (i.e., belonging to the same ensemble) action in the crowd. Furthermore, in order to offer the possibility to "weight" the segmentation considering also the number of people and their proximity[7],

[6] You can find the annotation manual equipped with visual examples of each atomic component at the following address: http://mmlab.disi.unitn.it/extra/oz/.

[7] If A is engaged in an action of the satisfaction ensemble, and B, next (or near) to the former in the stands, is engaged in an action of the dissatisfaction, or the disengagement ensembles, we can fairly assume they do not belong to the same fan group. If instead B is engaged in cheering—depending on how many people around A and B are doing something labeled as cheering or satisfaction —often, they belong to the same fan group.

Table 7.8 Annotation manual

Sub-set (variable)	Definition	Possible choices (variable's states)
Position	The spectator's full body position. Sitting and standing cannot occur simultaneously. If the person is nor sitting nor standing, see Locomotion	Sitting
		Standing
		Locomotion (cf. further)
Posture: superior half	The spectator's posture: this refers to the posture of different body parts (except head—cf. Head Pose). Postures of the same body part cannot occur simultaneously; postures of different body parts *do* occur simultaneously	Arms alongside body
		Crossed arms
		Elbows (or forearms) on legs
		Hands on hips
		Hands on legs
		Hand in one's womb and/or Joined/ crossed hands
		Hands in pocket
Posture: inferior half		Crossed legs or ankle on knee
		Parallel legs (straight or bended)
Locomotion	The spectator's horizontal and vertical locomotion	Walking
	N.B. With respect to jumping: annotate *each* jump (the highest point)	Jumping (*annotate each jump*)
		Rising pelvis slightly up
Action	The spectator activity. This refers to different body parts. Actions of the same body part cannot occur simultaneously; actions of different body parts *do* occur simultaneously. N.B. With respect to clapping: annotate *each* clap	Pointing toward game field
		Pointing toward something which is not on the game field
		Rising and keeping arms over head
		Waving arms
		Shaking flag or another "fan-object"
		Bringing hands in cone around mouth
		Whistling
		Producing a "positive" iconic gesture (e.g. victory gesture)
		Producing a "negative" iconic gesture (e.g. flipping off)
		Applauding
		Clapping, *i.e.* beating one's hand: a. against the other one; b. against another body part; c. against an object (*annotate each clap*)
		Using camera/phone to take a photo/ video
		Using binoculars
		Using megaphone
		Patting on another person's back or shoulder or tight to cheer
		Touch a person to get his attention
		Hugging another person
		Kissing another person

Table 7.8 (continued)

Sub-set (variable)	Definition	Possible choices (variable's states)
		Passing an object to another person
		Hitting/beating/punching another person for fun
		Hitting/beating/punching another person for real
		Bringing hand to forehead, or to cheeks, or to mouth
		Hitting hand/s on tight/s (once)
		Opening arms
Person's bounding box	With the term "bounding box," we refer to the maximum area visible for each subject (thus not including, for example, occluded legs)	Coordinates are defined by drawing a rectangular box around the subject of interest. This is referred to the parameter FULL BODY
Person's head box	Similarly as before, but focusing only on the head area	Coordinates are defined by drawing a rectangular box around the head of the subject of interest. This is referred to the parameter HEAD
Person's head pose	Head pose identifies where the person is looking at as seen by the camera. For example *looking left*, implies the user is *looking* at the left side of your screen	Frontal (to game field)
		Half-left (to game field)
		Half-right (to game field)
		Left, Right, or Back (away from game field)
		Down (or any other option different from the above mentioned ones)
Head visible	Is the head of the person clearly visible? About 50 % is at least required	True
		False

a differential index between ensembles, named *Ensembles Differential index* (EDi) has been elaborated (Table 7.10). As a further method to test and compare with the former, a similarity index between ensembles' components, called *Action Similarity Index* (ASi), has been also created (see in Fig. 7.3 a portion of the matrix).

2. With respect to the tasks of attention and excitement level calculation (within-individual analysis), two indexes for, respectively, attention and excitement—where the former parallels reading the field, and the latter performing the stands—have been elaborated. Called *Attention index* (Ai) and *Excitement index* (Ei), they must be regarded as relative rather than absolute indexes, and they are processual. I do not assign a value to the level of attention/excitement *per se*, I merely point out that, for instance, when "waving arms" is present at t1 and was not at t0, then we can assume that Ei has increased. Therefore, to take into consideration the processual and sequential character of action-in-interaction, *intra-variable dynamic values* for both attention and excitement have been assigned. They mark a de/increase (or not) in the level of attention/excitement whenever

Table 7.9 Action ensembles (in green: components belonging to more than one ensemble)

Engagement Ensemble	Disengagement Ensemble	Cheering Ensemble	Satisfaction Ensemble	Dissatisfaction Ensemble
Standing	Sitting	Jumping	Jumping	"Negative" iconic gesture
Rising pelvis slightly up	Crossed arms	Waving arms	Rising and keeping arms over head	Bringing hands to forehead/mouth
Elbows/forearms on legs	Hands on hips	Shaking flag or other fan-object	"Positive" iconic gesture	Hitting hand/s on tight/s (once)
	Hands on legs	Whistling	Applauding	Opening arms
	Hands in womb and/or joined	Clapping	Clapping	
	Hands in pockets			

Table 7.10 Ensembles Differential index (EDi)

	Engagement	Disengagement	Cheering	Satisfaction	Dissatisfaction
Engagement	0	1	1	2	1
Disengagement		0	2	2	1
Cheering			0	1	1
Satisfaction				0	2
Dissatisfaction					0

a variable changes state. As you can see from a portion of the matrix in Fig. 7.4, for each sub-set of components (variable), a value (−1, 0 or +1) is assigned, on both attention (dynA) and excitement (dynE), to the passage from one to another component of that sub-set (state). This can be computed sequentially (Markov model) and compositionally (sum of all the sub-sets' dynamic values at any given point in time), as exemplified in Fig. 7.5. Four specifications are needed. First, Locomotion and Action sub-sets also have a "none" state. Second, they contain one component each—[jumping] and [clapping], respectively—for which a value (dynE+1) is assigned to intra-state "change" (e.g., [jump] to [jump]), since they are regarded as "crescendo activities," so to speak. Third, the Action sub-set is not a variable with mutually excluding states; therefore, the model foresees the possibility of none, one, or more actions at the same point in time. Fourth, the Arms&Hands sub-set has not a "none" state (since arms/hands need to be kept in one posture or the other), but can have a "blank" state ([/]) when particular actions are performed (e.g., [waving arms]). Indeed, in order to check the coherence of the model, and to better consider all sub-sets altogether (the individual's full body and conduct as a whole), an *inter-variable compositional matrix* has been created. The latter allows to exclude physically impossible combinations (e.g., [sitting] + [walking], or [hands on hips] + [waving arms]), and bodily "weird" and/or rare ones (e.g., [walking] + [hands on legs], or [jumping] + [hands joined]).

	Standing	Rising pelvis slightly up	Elbows/ forearms on legs	Sitting	Crossed arms	Hands on hips	Hands on legs	Hands in womb and/ or joined	Hands in pockets	Jumping
Standing	+2	+1	+1	0	+1	+1	0	0	+1	+1
Rising pelvis slightly up		+2	+1	+1	0	0	0	0	0	0
Elbows/forearms on legs			+2	+1	0	0	0	0	0	0
Sitting				+2	+1	0	+1	+1	+1	0
Crossed arms					+2	+1	+1	+2	0	0
Hands on hips						+2	+1	+1	+1	0
Hands on legs							+2	+1	+1	0
Hands in womb and/or joined								+2	+1	0
Hands in pockets									+2	0
Jumping										+2

Fig. 7.3 Portion of the matrix of the Action Similarity index (ASi)

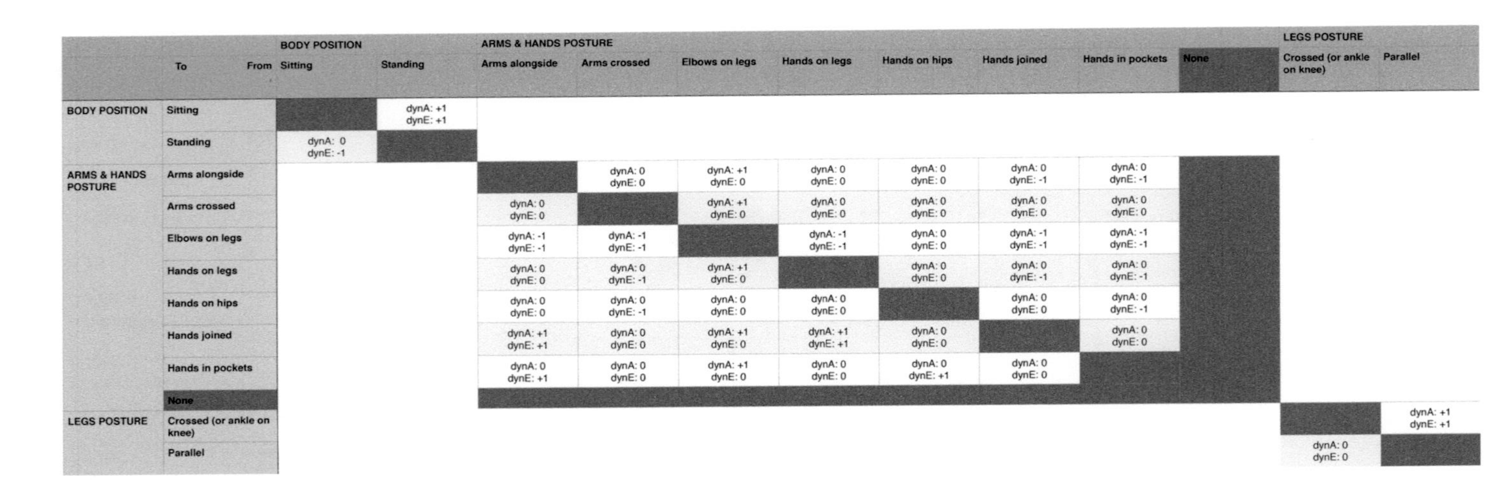

	To / From	BODY POSITION: Sitting	BODY POSITION: Standing	ARMS & HANDS POSTURE: Arms alongside	Arms crossed	Elbows on legs	Hands on legs	Hands on hips	Hands joined	Hands in pockets	None	LEGS POSTURE: Crossed (or ankle on knee)	Parallel
BODY POSITION	Sitting		dynA: +1 dynE: +1										
	Standing	dynA: 0 dynE: -1											
ARMS & HANDS POSTURE	Arms alongside				dynA: 0 dynE: 0	dynA: +1 dynE: 0	dynA: 0 dynE: 0	dynA: 0 dynE: 0	dynA: 0 dynE: -1	dynA: 0 dynE: -1			
	Arms crossed			dynA: 0 dynE: 0		dynA: +1 dynE: 0	dynA: 0 dynE: 0	dynA: 0 dynE: 0	dynA: 0 dynE: 0	dynA: 0 dynE: 0			
	Elbows on legs			dynA: -1 dynE: -1	dynA: -1 dynE: -1		dynA: -1 dynE: -1	dynA: 0 dynE: 0	dynA: -1 dynE: -1	dynA: -1 dynE: -1			
	Hands on legs			dynA: 0 dynE: 0	dynA: 0 dynE: -1	dynA: +1 dynE: 0		dynA: 0 dynE: 0	dynA: 0 dynE: -1	dynA: 0 dynE: -1			
	Hands on hips			dynA: 0 dynE: 0	dynA: 0 dynE: -1	dynA: 0 dynE: 0	dynA: 0 dynE: 0		dynA: 0 dynE: 0	dynA: 0 dynE: -1			
	Hands joined			dynA: +1 dynE: +1	dynA: 0 dynE: 0	dynA: +1 dynE: 0	dynA: +1 dynE: +1	dynA: 0 dynE: 0		dynA: 0 dynE: 0			
	Hands in pockets			dynA: 0 dynE: +1	dynA: 0 dynE: 0	dynA: +1 dynE: 0	dynA: 0 dynE: 0	dynA: 0 dynE: +1	dynA: 0 dynE: 0				
	None												
LEGS POSTURE	Crossed (or ankle on knee)												dynA: +1 dynE: +1
	Parallel											dynA: 0 dynE: 0	

Fig. 7.4 Portion of the matrix for intra-variable dynamic Attention and Excitement. (dynA, dynE)

	t0		t1		t2		t3		t4
Body Position	[sitting]	→ A0 E0	[sitting]	→ A+1 E+1	[standing]	→ A0 E0	[standing]	→ A0 E0	[standing]
Posture: superior	[hands joined]	→ A0 E0	[hands joined]	→ A+1 E+1	[hand on hips]	→ A0 E0	[arms alongside]	→ A0 E0	[/]*
Posture: inferior	[crossed legs]	→ A+1 E+1	[parallel legs]	→ A0 E0	[parallel legs]	→ A0 E0	[parallel legs]	→ A0 E0	[parallel legs]
Locomotion	[none]	→ A+1 E+1	[pelvis up]	→ A-1 E-1	[none]	→ A0 E+1	[jumping]	→ A0 E+1	[jumping]
Action 1	[none]	→ A0 E0	[none]	→ A0 E0	[none]	→ A0 E0	[none]	→ A0 E+1	[raising arms]*
Action 2	[none]	→ A0 E0	[none]	→ A0 E0	[none]	→ A0 E0	[none]	→ A0 E0	[none]
Head Pose	[frontal]	→ A0 E0	[half left]	→ A0 E0	[half left]	→ A0 E0	[half left]	→ A0 E0	[frontal]
dynA, dynE		**A+2 E+2**		**A+1 E+1**		**A0 E+1**		**A0 E+2**	
Attention index	A0		A2		A3		A3		A3
Excitement index	E0		E2		E3		E4		E6

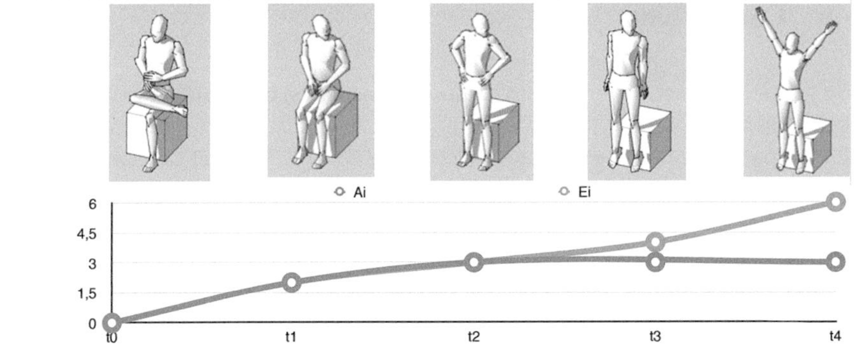

Fig. 7.5 An example of sequential and compositional analysis

3. With respect to the event segmentation task, alongside aggregating individual ongoing levels of attention (Ai) and excitement (Ei) and looking for positive peaks (especially in aggregated excitement, aggEi), a *Satisfaction index* (Si) has been elaborated, marking the ongoing level of enjoyment/satisfaction of/for game actions/' outcome. Only in presence of dynE + 1 a dynamic value can be assigned to enjoyment/satisfaction (dynS). Yet this is a necessary but insufficient condition for determining the value. Consider, for instance, kissing: it marks an increase in the excitement level, but it can fairly have nothing to do with satisfaction w.r.t. the game. This means that dynS can be not applicable (n.a. = 0), and, when applicable, can be determinable (− 1, + 1) or not (n.d. = 0). By aggregating individual Si and looking for both positive and negative peaks in aggregated satisfaction (aggSi), one obtains a method for event segmentation other than that based on positive aggEi peaks, and one can:
   - compare the two methods and see which one performs better in terms of accuracy as well as required time and computational effort;
   - modulate event segmentation on the basis of the purpose at hand (e.g. shorter or longer report) by using either aggEi-based or aggSi-based methods.

Moreover, the Si can be used to test and/or enhance crowd segmentation (cf. task 1) by comparing EDi-based and ASi-based methods, on the one hand, with, on the other hand, a method based instead on the distance between co-occurrent peaks in each and every couple of mutually proximal persons' Si "flowcharts". Since a positive value points to satisfaction and a negative one to dissatisfaction, looking for the maximum distance should help in "drawing the line," so to speak, and identifying different fan groups.

As previously mentioned, we are now concluding the axiomatization and we shall soon test the related algorithms. We expect such a sociologically driven, ontologically-founded automated analysis to outperform previous Social Signal Processing analyses.

7.6 Conclusions

We live in complex, variously interconnected societies; in technologically dense as well as demographically packed environments; in *systems* that yet are (re)produced during and through our everyday situated interactions. To meet contemporary societal challenges, therefore, we need to deal with socio-technical complexity at a systemic yet micro-grounded level. The above illustrated approach is intended to take up such a challenge. Besides its application to the considered project/scenario and the related specific objectives and results (cf. Conigliaro et al. 2013a, b, 2015), it presents several advantages, we believe.

From a scientific point of view,

1. it can be fruitfully applied to other datasets concerning spectatorship, thus allowing comparative analysis (e.g., fans of different sports where spectators are differently arranged and the arena differently designed, but also, with few modifications, movie vs. theatre attendees, etc.);
2. it opens the road for a substantial updating in computer vision methods, especially for semi-supervised "crowd behavior" analysis (cf. Conigliaro et al. 2015);
3. it allows both domain and cross-domain ontology learning and updating, thanks to the recursive test of the model in (diverse) empirical contexts;
4. it makes feasible, at last, to conduct *micro*-sociology on a *large* scale, leveraging on the automated yet detailed and sociologically meaningful analysis that the method eventually allows.

From the perspective of concrete applications, on the other hand, consider just a couple of examples:

- augmented video-summarization: the spectators feedback, automatically recognized, may help in highlighting exciting or crucial events that should be included in a video summarization of the show/event;
- augmented monitoring/video-surveillance: discriminating, in order to foresee subsequent people behavior, whether, for instance, a display of excitement is determined by a rejoicing attitude or not, or, more generally, whether it is related to—and "explainable" by, i.e., *accountable* through (Garfinkel 1967)—the

activity at hand or not (e.g., whether it co-occurs with a goal or not, whether the participants' focus of attention rests on the game-field or elsewhere, whether and how much the detected display is mutually coordinated with the surrounding participants' equally-detected actions).

I believe we opened a promising, interdisciplinary research avenue. We will pursue that in the attempt to extend the approach to even more complex scenarios, such as the airport of the VisCoSo project. I hope others will join in this scientific endeavor.

Acknowledgments Besides all members of the OZ project, I wish to thank Marco Cristani, Roberta Ferrario and Daniele Porello for the helpful discussions about compositional methods to be axiomatized and computed.

Funding Acknowledgments This work is part of the OZ project, financed by the Winter Universiade Trentino 2013 Educational Programme. Chiara Bassetti is supported by the VisCoSo project grant, funded by the Autonomous Province of Trento through the "Team 2011" programme.

References

Allison, M. (1979). The game: A participant observation study. *Journal of Sport Behavior, 2,* 93–102.

Argyle, M., & Dean, J. (1965). Eye-contact, distance and affiliation. *Sociometry, 28*(3), 289–304.

Atkinson, M. (1984). *Our masters' voices: The language and body language of politics*. New York: Routledge.

Avgerou, C., Ciborra, C., & Land, F. F. (Eds.). (2004). *The social study of information and communication technology: Innovation, actors and context.* Oxford: Oxford University Press.

Bassetti, C., Bottazzi, E., & Ferrario, R. (2013). Fatal attraction. Interaction and crisis management in socio-technical systems. 29th EGOS Colloquium (European Group for Organizational Studies), Montreal, July 4–6, 2013.

Bateson, G. (1968). Redundancy and coding. In T. A. Sebeok (Ed.), *Animal communication: Techniques of study and results of research* (pp. 614–626). Bloomington: Indiana University Press.

Baxter, J. C., & Rozelle, R. M. (1975). Nonverbal expression as a function of crowding during a simulated police-citizen encounter. *Journal of Personality and Social Psychology, 32*(1), 40–54.

Berlonghi, A. (1995). Understanding and planning for different spectator crowds. *Safety Science, 18,* 239–247.

Birdwhistell, R. L. (1952). *Introduction to kinesics: An annotation system for analysis of body motion and gesture*. Louisville: University of Louisville.

Birdwhistell, R. (1970). *Kinesics and context: Essays on body motion communication*. Philadelphia: University of Pennsylvania Press.

Blumer, H. (1951). Collective behavior. In A. McClung Lee (Ed.), *Principles of sociology*. New York: Barnes & Noble.

Blumer, H. (1969). *Symbolic interactionism: Perspective and method.* Englewood Cliffs: Prentice-Hall.

Bowker, A., Boekhoven, B., Nolan, A., Bauhaus, S., Glover, P., Powell, T., & Taylor, S. (2009). Naturalistic observations of spectator behavior at youth hockey games. *Sport Psychologist, 23,* 301–316.

Burgoon, J. K., Buller, D. B., & Woodall W. G. (1989). *Nonverbal communication: The unspoken dialogue*. New York: Harper & Row.

Button, G., & Sharrock, W. (1998). The organizational accountability of technological work. *Social Studies of Science, 28*(1), 73–102.

Choi, Y. S., Martin J. J., Park, M., & Yoh, T. (2009). Motivational factors influencing sport spectator involvement at NCAA Division II basketball games. *Journal for the Study of Sports and Athletes in Education, 3*(3), 265–284.

Ciolek, T. M., & Kendon, A. (1980). Environment and the spatial arrangement of conversational encounters. *Sociological Inquiry, 50,* 237–271.

Collins, R. (1988). Theoretical continuities in Goffman's work. In P. Drew & A. Wootton (Ed.), *Erving Goffman. Exploring the interaction order* (pp. 41–63). Oxford: Polity Press.

Conigliaro, D., Setti, F., Bassetti, C., Ferrario, R., & Cristani, M. (2013a). Attento: Attention observed for automated spectator crowd monitoring. In A. A. Salah, H. Hung, O. Aran, & H. Gunes (Eds.), *Human behavior understanding*, series *Lecture notes in computer science* (Vol. 8212, pp. 102–111). Heidelberg: Springer.

Conigliaro, D., Setti, F., Bassetti, C., Ferrario, R., & Cristani, M. (2013b). Viewing the viewers: A novel challenge for automated crowd analysis. *International Workshop on Social Behavior Analysis* (SBA2013, in conjunction with ICIAP), Naples, September 11–13, 2013.

Conigliaro, D., Setti, F., Bassetti, C., Ferrario, R., & Cristani, M. (2015). The S-HOCK dataset: Analyzing crowds at the stadium. IEEE Conference on Computer Vision and Pattern Recognition (CVPR), Boston, MA (USA), 7–12 June 2015 (pp. 2039–2047). doi: 10.1109/CVPR.2015.7298815. http://ieeexplore.ieee.org/xpl/articleDetails.jsp?arnumber=7298815&queryText=cvpr%202015%20s-hock&newsearch=true.

Cristani, M., Murino, V., & Vinciarelli, A. (2010). Socially intelligent surveillance and monitoring: Analysing social dimensions of physical space. In IEEE Computer Society Conference on Computer Vision and Pattern Recognition Workshops (CVPRW) (pp. 51–58). doi:10.1109/CVPRW. 2010.5543179.

de Waal, F. B. M., Aureli, F., & Judge, P. G. (2000). Coping with crowding. *Scientific American, 282*(5), 76–81.

Dittman, A. T. (1987). The role of body movement in communication. In A. W. Siegman & S. Feldstein (Eds.), *Nonverbal behavior and communication*. Hillsdale: Lawrence Erlbaum.

Doermann, D., & Mihalcik, D. (2000). Tools and techniques for video performance evaluation. IEEE Proceedings of the 15th International Conference on Pattern Recognition (Vol. 4, pp. 167–170). doi:10.1109/ICPR.2000.902888.

Duncan, S., & Fiske, D. W. (1977). *Face-to-face interaction: Research, methods, and theory*. Hillsdale: Hillsdale: Erlbaum.

Durkheim, E. (1965). *The elementary forms of the religious life*. New York: Free Press.

Edwards, J. (1979). The home-field advantage. In J. H. Goldstein (Ed.), *Sports, games, and play: Social and psychological viewpoints* (pp. 409–430). Hillsdale: Lawrence Erlbaum.

Ekman, P., & Friesen W. V. (1969a). Nonverbal leakage and clues to deception. *Psychiatry, 32*(1), 88–106

Ekman, P., & Friesen, W. V. (1969b). The repertoire of nonverbal behavior. Categories, origins, usage, and coding. *Semiotica, 1,* 49–98.

Ekman, P., & Friesen, W. V. (1972). Hand movements. *Journal of Communication, 22,* 353–374.

Ferrario, R. (2011). VisCoSo Project Proposal. http://www.loa.istc.cnr.it/projects/viscoso/proposal.

Garfinkel, H. (1967). *Studies in ethnomethodology*. Englewood Cliffs: Prentice-Hall.

Garfinkel, H. (2002). *Ethnomethodology's program: Working out Durkeim's aphorism*. Lanham: Rowman & Littlefield Publishers

Garfinkel, H. (2006). *Seeing sociologically: The routine grounds of social action*. Boulder: Paradigm Publishers.

Givens D. B. (2002). *The non-verbal dictionary of gestures, signs & body language cues*. Washington, DC: Center for Nonverbal Studies Press.

Goffman, E. (1959). *The presentation of self in everyday life*. New York: Anchor Books.

Goffman, E. (1961). *Encounters: Two studies in the sociology of interaction*. Indianapolis: Bobbs-Merrill.

Goffman, E. (1963). *Behavior in public places: Notes on the social organization of gatherings*. New York: Free Press of Glencoe.
Goffman, E. (1967). *Interaction ritual: Essays on face-to-face behavior*. Chicago: Aldine Publishing.
Goffman, E. (1981). *Forms of talk*. Philadelphia: University of Pennsylvania Press.
Goode, E. (1992). *Collective behavior*. Fort Worth: Saunders College Pub.
Goldstein, J. H., & Arms, R. L. (1971). Effects of observing athletic contests on hostility. *Sociometry, 34*(1), 83–90.
Grusky, O. (1963). The effects of formal structure on managerial recruitment: A study of baseball organization. *Sociometry, 26*(3), 345–353.
Guarino, N. (1998). Formal ontology and information systems. In N. Guarino (Ed.), Formal Ontology in Information Systems. Proceedings of FOIS'1998, Trento, Italy, June 6–8 (pp. 3–15). Amsterdam: IOS Press.
Guarino, N. (2009). The ontological level: Revisiting 30 years of knowledge representation. In A. Borgida, V. Chaudhri, P. Giorgini, & E. Yu (Eds.), *Conceptual modelling: Foundations and applications* (pp. 52–67). Berlin: Springer.
Guillaumin, C. (1992). *Le corps construit, in Guillaumin Sexe, race et pratique du pouvoir: l'idée de Nature* (pp. 7–42). Paris: Coté-femmes.
Hall, E. T. (1963). A system for the notation of proxemic behavior. *American Anthropologist, 65*(5), 1003–1026.
Hall, E. T. (1966). *The hidden dimension*. Garden City: Anchor Books.
Heath, C., Hindmarsh, J., & Luff, P. (2010). *Video in qualitative research. Analysing social interaction in everyday life*. London: Sage.
Heritage, J. (1999). Conversation analysis at century's end: Practices of talk-in-interaction, their distributions, and their outcomes. *Research on Language and Social Interaction, 32,* 1–2.
Heritage, J., & Greatbatch, D. (1986). Generating applause: A study of rhetoric and response at party political conferences. *American Journal of Sociology, 92*(1), 110–157.
Hewes, G. W. (1957). The anthropology of posture. *Scientific American, 196*(2), 122–132.
Hietanen, J. K. (2002). Social attention orienting integrates visual information from head and body orientation. *Psychological Research, 66*(3), 174–179.
Hocking , J. E., Margreiter, D. G., & Hylton, C. (1977). Intra-audience effects: A field test. *Human Communication Research, 3*(3), 243–249.
Hocking, J. E. (1982). Sports and spectators: Intra-audience effects. *Journal of Communication, 32*(1), 100–108.
Hylton, C. G. (1971). The effects of observable audience response on attitude change and source credibility. *Journal of Communication, 21,* 253–265.
James, W. T. (1932). A study of the expression of bodily posture. *Journal of General Psychology, 7,* 405–437.
Kendon, A. (1967). Some functions of gaze-direction in social interaction. *Acta Psychologica, 26,* 22–63.
Kendon, A. (1970). Movement coordination in social interaction: Some examples described. *Acta Psychologica, 32,* 101–125.
Kendon, A. (1972). Some relationships between body motion and speech. In A. Seigman & B. Pope (Eds.), *Studies in dyadic communication* (pp. 177–216). Elmsford: Pergamon Press.
Kendon, A. (1990). *Conducting interaction: Patterns of behavior in focused encounters*. Cambridge University Press.
Kendon, A. (1992). The negotiation of context in face-to-face interaction. In A. Duranti & C. Goodwin (Eds.), *Rethinking context: Language as an interactive phenomenon* (pp. 323–334). Cambridge: Cambridge University Press.
Kendon, A. (2004). *Gesture: Visible action as utterance*. Cambridge: Cambridge University Press.
Kerstetter, D., & Kovich, G. M. (1997). The involvement profiles of division I women's basketball spectators. *Journal of Sport Management, 11,* 234–249.
Knapp, M. L. (1972). *Nonverbal communication in human interaction*. New York: Holt, Rinehart and Winston.

Knapp, M. L. (1978). The effect of the eye behavior on human communication. In M. L. Knapp (Ed.), *Nonverbal communication in human interaction* (2nd ed., pp. 294–321). New York: Holt, Rinehard and Winston.
Lafrance, M., & Broadbent, M. (1976). Group rapport: Posture sharing as a nonverbal indicator. *Group and Organization Management, 1*(3), 328–333.
Lefebvre, L. M. (1975). Encoding and decoding of ingratiation in modes of smiling and gaze. *The British Journal of Social and Clinical Psychology, 14*(1), 33–42.
Levy, L. (1989). A study of sports crowd behavior: The case of the great pumpkin incident. *Journal of Sport and Social Issues, 13*(2), 69–91.
Madrigal, R. (1995). Cognitive and affective determinants of fan satisfaction with sporting event attendance. *Journal of Leisure Research, 27*(3), 205–207.
Madrigal, R., & James, J. (1999). Team quality and the home advantage. *Journal of Sport Behavior, 22,* 381–398.
Mann, L. (1979). Sports crowds and the collective behavior perspective. In J. Goldstein (Ed.), *Sport, games and play: Social and psychological viewpoints* (pp. 229–327). Hillsdale: Laurence Earlbaum.
Matsumoto, D., & Kudoh, T. (1987). Cultural similarities and differences in the semantic dimensions of body postures. *Journal of Nonverbal Behavior, 11,* 166–179.
McDonald, M. A., Milne, G. R., & Hong, J. (2002). Motivational factors for evaluating sport spectator and participant markets. *Sport Marketing Quarterly, 11,* 100–113.
McNeill, D. (2005). *Gesture and thought*. Chicago: University of Chicago Press.
McPhail, C. (1991). *The myth of the madding crowd*. New York: De Gruyter.
McPhail, C. (1994). From clusters to arcs and rings. *Research in Community Sociology, 1,* 35–57.
Mead, G. H. (1934). *Mind, self and society: From the stand-point of a social behaviorist*. Chicago: University of Chicago Press.
Mehrabian, A. (1969). Significance of posture and position in the communication of attitude and status relationships. *Psychological Bulletin, 71*(5), 359–372.
Mehrabian, A. (1972). *Nonverbal communication*. Chicago: Aldine-Atherton.
Mondada, L. (2008). Doing video for a sequential and multimodal analysis of social interaction: Videotaping institutional telephone calls. *FQS—Forum: Qualitative Sozialforschung/Forum: Qualitative Social Research*, 9(3). http://www.qualitative-research.net/index.php/fqs/article/view/1161.
Morris, D. (1994). *Bodytalk: The meaning of human gestures*. New York: Crown Trade Paperbacks.
Nielsen, G. (1962). *Studies in self-confrontation*. Copenhagen: Monksgaard.
Park, R. E., Burgess, E. W., & McKenzie, R. D. (1925). *The city*. Chicago: The University of Chicago Press.
Psathas, G. (Ed.). (1995). *Conversation analysis: The study of talk-in-interaction*. USA: Sage.
Rautaray, S. S., & Agrawal, A. (2015). Vision based hand gesture recognition for human computer interaction: A survey. *Artificial Intelligence Review, 43,* 1–54.
Rawls, A. W. (2011). Wittgenstein, Durkheim, Garfinkel and Winch: Constitutive orders of sense-making. *Journal for the Theory of Social Behaviour, 41*(4), 396–418.
Roadburg, A. (1980). Factors precipitating fan violence: A comparison of professional soccer in Britain and North America. *The British Journal of Sociology, 31*(2), 265–276.
Rosenfeld, L. B. (1973). *Human interaction in the small group setting*. Columbus: Merrill.
Sacks, H. (1992). *Lectures on conversation*. Oxford: Blackwell.
Sapir, E. (1927). Language as a form of human behavior. *The English Journal, 16*(6), 421–433.
Schechner, R. (1971). Audience participation. *The Drama Review: TDR, 15*(3), 73–89.
Schechner, R. (1986). *Performance theory*. New York: Routledge.
Scheflen, A. E. (1964). The significance of posture in communication systems. *Psychiatry, 27*(4), 316–331.
Scheflen, A. E. (1972). *Body language and the social order*. Englewood Cliffs: Prentice Hall.
Schwartz, B., & Barsky, S. F. (1977). The home advantage. *Social Forces, 55,* 641–661.

Schweingruber, D. & McPhail, C. (1999). A method for systematically observing and recording collective action. *Sociological Methods & Research, 27*(4), 451–498.

Setti, F., Russel, C., Bassetti, C., & Cristani, M. (2015). F-formation detection: Individuating free-standing conversational Groups in images. *PLOS One, 10*(5), e0123783.

Simmel, G. (1908). *Soziologie*. Leipzig: Duncker & Humblot.

Sloan, L. (1979). The motives of sports fans. In J. Goldstein (Ed.), *Sport, games and play: Social and psychological viewpoints* (pp. 175–226). Hillsdale: Laurence Earlbaum.

Suchman, L. (2002). Practice-based design of information systems: Notes from an hyperdeveloped world. *The Information Society, 8*(2), 139–144.

Vinciarelli, A., Pantic, M., & Bourlard, H. (2009). Social signal processing: Survey of an emerging domain. *Image and Vision Computing Journal, 27*, 1743–1759.

Vrugt, A., & Kerkstra, A. (1984). Sex differences in nonverbal communication. *Semiotica,* 50-(1–2), 1–42.

Whitworth, B. (2009). A brief introduction to sociotechnical systems. In M. Khosrow-Pour (Ed.), *Encyclopedia of information science and technology* (2nd ed., pp. 394–400). Hershey: IGI Global.

Yiannakis, A., Melnick, M. J., & McIntyre, T. D. (1993). *Sport sociology: Contemporary themes*. Dubuque: Kendall/Hunt Pub. Co.

Young, I. M. (1980). Throwing like a girl: A phenomenology of feminine body comportment motility and spatiality. *Human Studies, 3*(2), 137–156.

Zillman, D., Bryant, J., & Sapolsky, N. (1989). Enjoyment from sports spectatorship. In J. Goldstein (Ed.), *Sport, games and play* (pp. 241–278). Hillsdale: Lawrence Earlbaum.

Chapter 8
Trends in Social Science: The Impact of Computational and Simulative Models

LABSS-ISTC-CNR

Rosaria Conte, Mario Paolucci, Stefano Picascia and Federico Cecconi

8.1 The Survey: Method and Caveats

The simple method that we used is based on a survey conducted with the help of Google Scholar (Beel 2009; Chen 2010) comparing the growth (or decrease) of publications' citations for different disciplines in the social and computational sciences in the last decade with the relative growth of agent-based simulation and social simulation (Squazzoni 2008, 2010) within each discipline. This is quite a simple approach, indeed, which tells us nothing about progress in research quality or achieved breakthroughs but may help us compare the expansion pace of different research fields.

We performed search-engine queries using one "computational/simulation" tag and one discipline label. The queries were expressed in the form *tag + discipline label + year*. For each query, the raw citation number for every year ranging 2006 to 2011 was recorded. Subsequently, the variation between the years was calculated and the variation in the number of citations in each discipline/tag records compared.

We used eight tags:

- Agent-based simulation
- Monte carlo simulation
- Network analysis
- Neural network
- Numerical simulation

R. Conte (✉) · M. Paolucci
Institute of Cognitive Sciences and Technologies, CNR LABSS. Laboratory of Agent-Based Social Simulation, Via Palestro 32, 00185 Rome, Italy
e-mail: rosaria.conte@istc.cnr.it

M. Paolucci
e-mail: mario.paolucci@istc.cnr.it

F. Cecconi
Institute of Cognitive Science and Technology – CNR, Rome, Italy

F. Cecconi (ed.), *New Frontiers in the Study of Social Phenomena*,
DOI 10.1007/978-3-319-23938-5_8

- Reinforcement learning
- Game theory
- System dynamic

The first six tags are computational tags. *Agent-based simulation* identifies the class of simulation using artificial agents interacting with one another. *Monte carlo simulation* refers to the generic class of stochastic simulation models. *Network analysis* indicates the set of models that refers to complex network environment. *Neural network*, *numerical simulation,* and *reinforcement learning* refer to computational frameworks typically applied to the study of learning and adaptation in dynamic environments. *Game theory* and *system dynamics* are not *strictly* computational tags that we inserted in our dataset for the sake of comparison.

To effectuate our search, we used to Google Scholar (GS), mainly for convenience and ease of access. Since its introduction in 2005, GS has elicited mixed feelings in the scientific community. The first research papers reporting on its coverage found GS wanting (Neuhaus and McCulloch 2006). However, as was to be expected, Google improved and enlarged its coverage, and the current literature (Chen 2010) reports that sources that in 2005 had low coverage (ranging from 30 to 88 %) have now reached between 98 and 100 % coverage. In addition, GS is known to index sources, such as conference proceedings, working papers, and technical reports usually not included in other metrics (like ISI Web of Science).

It should be noted that our investigation suffers from some limitations, the most serious of which concerns the way we performed our queries, which can only detect literal matches but has no semantics. The text is searched as such in the articles' title and body, to ensure, for example, that a query for "Engineering social simulation" will not return articles *in* the engineering field, but articles generically referring *to* engineering. Nonetheless, GS's rough number of citations can be considered not only as a good proxy for the real number of occurrences of a particular keyword in scientific papers, but can be replaced with a more specific search in a yet-to-be-completed search engine.

Another caveat stems from the design of Google Scholar, which was not made with the purpose of retrieving large amounts of data, but to find specific papers—thus, we are stretching the tool's usage in a way that might cause some retrieval artifacts. To partially compensate, we ran the queries twice at two months' distance, finding substantial agreement.

8.2 Results

We analyzed seven disciplines—economics, history, philosophy, physics, psychology, sociology, and statistics. The criterion was to start a comparison between the traditional disciplines of nature (physics) and those of society (philosophy, economics, sociology, and psychology). Statistics and history were included to facilitate the understanding of the effect of some tags (see below). For each subject, we used

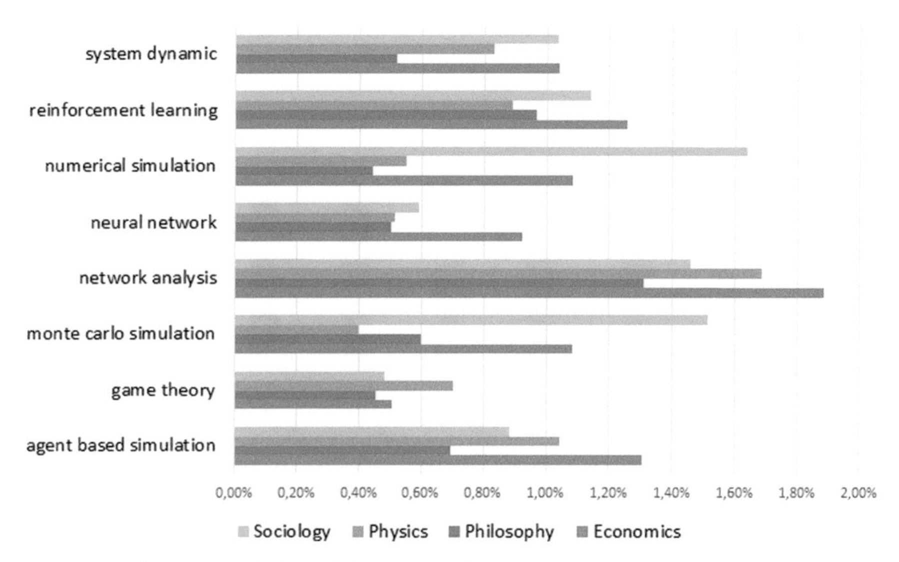

Fig. 8.1 The increase of citations during a 2-month period

eight different tags. We collected data at two different times separated by about two months. Figure 8.1 shows how the number of citations increased during the sampling period.

The increase shown in Fig. 8.1 was about 0.75 % on average, and this seems reasonable considering both the amount of scientific production and the indexing of further findings in these fields. Figure 8.2 shows the average (in percentage) of the differences of occurrences between the year 2005 and the listed year, up to 2011, for the various tags independent of disciplines.

Figures 8.3–8.6 show the trend of citations for economics, philosophy, psychology, and sociology (average on citations, Fig. 8.3), physics (Fig. 8.4), statistics (Fig. 8.5), and history (Fig. 8.6). All tags, in different measures, grow percentually in the number of citations. We can recognize a few overall trends: the first is the sharp increase in *network analysis* that dominates all the other ones through disciplines.

The second and third positions are occupied by the tags *agent-based simulation* and *reinforcement learning*, with *agent-based simulation* coming first in Figs. 8.4–8.6. The performance of the remaining tags depends on the discipline: In Fig. 8.3, numerical and Monte Carlo simulations come but game theory, which comes last also in history, but interestingly stays at the top of this last five tags group for physics. Monte Carlo simulation performs well in the social sciences (Fig. 8.3) but jumps between last and next-to-last position in all other ones. In statistics (Fig. 8.5), we found huge growth in *network analysis*. (note that we rescaled the y-axis of this graph, which has a maximum at 250 %, while the previous graphs had a maximum at 1808.4. %)

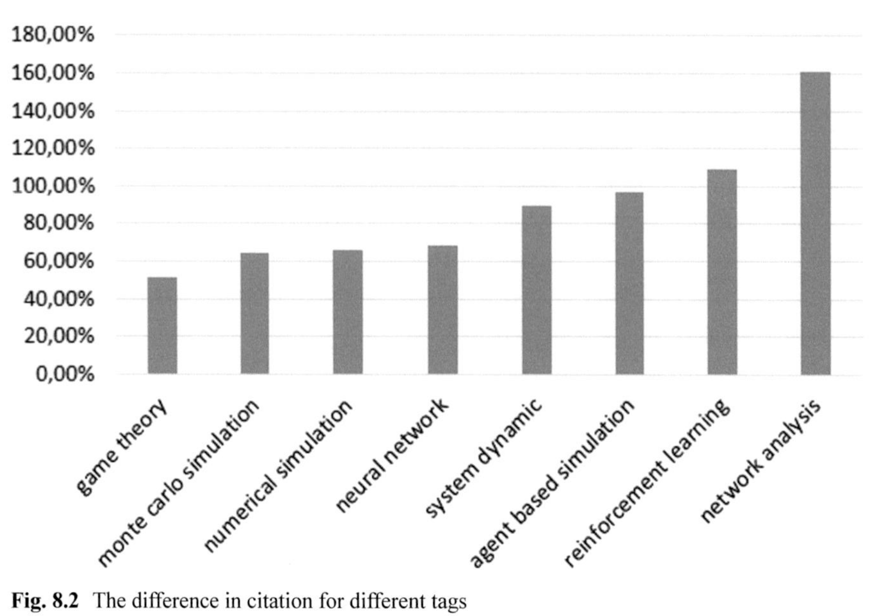

Fig. 8.2 The difference in citation for different tags

8.3 Discussion and Some Conclusions

Results show that some tags are more associated than others with the growth of the various disciplines (see Fig. 8.2). The difficult thing is to try to answer the question why. What does the clear affirmation of network analysis depend on? And why does

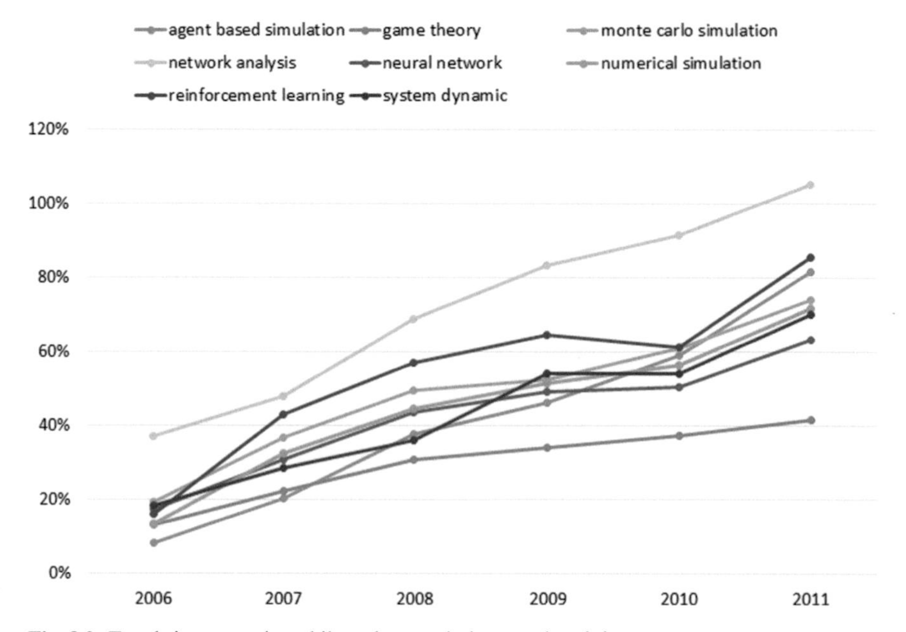

Fig. 8.3 Trends in economics, philosophy, psychology, and sociology

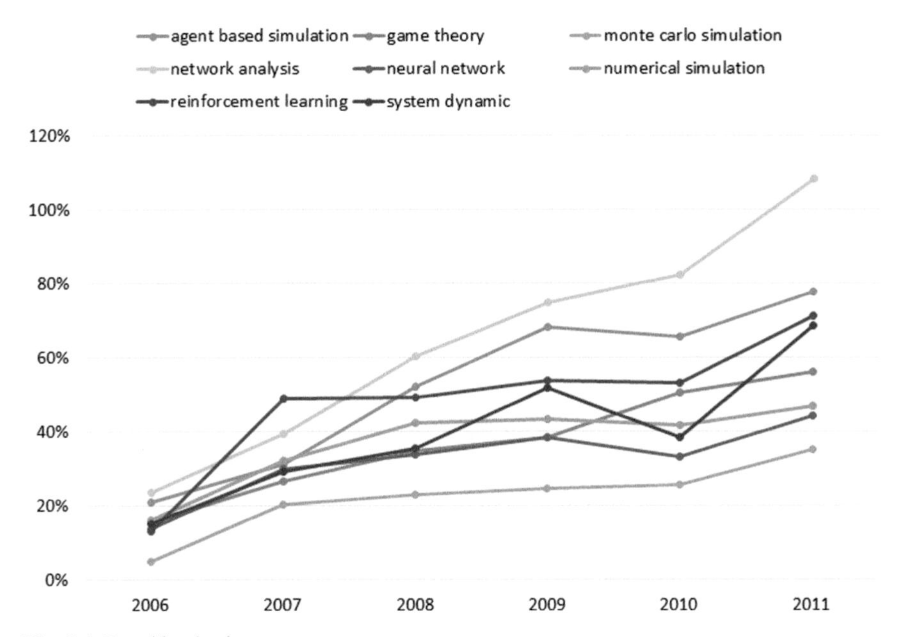

Fig. 8.4 Trend in physics

agent-based simulation, despite being relatively young compared, for example, to game theory, predominate in the disciplines of behavior that focus on these matters?

The larger increment is with the *network analysis* tag. With *network analysis* there is a convergence of interests on the part of disciplines, e.g., information and

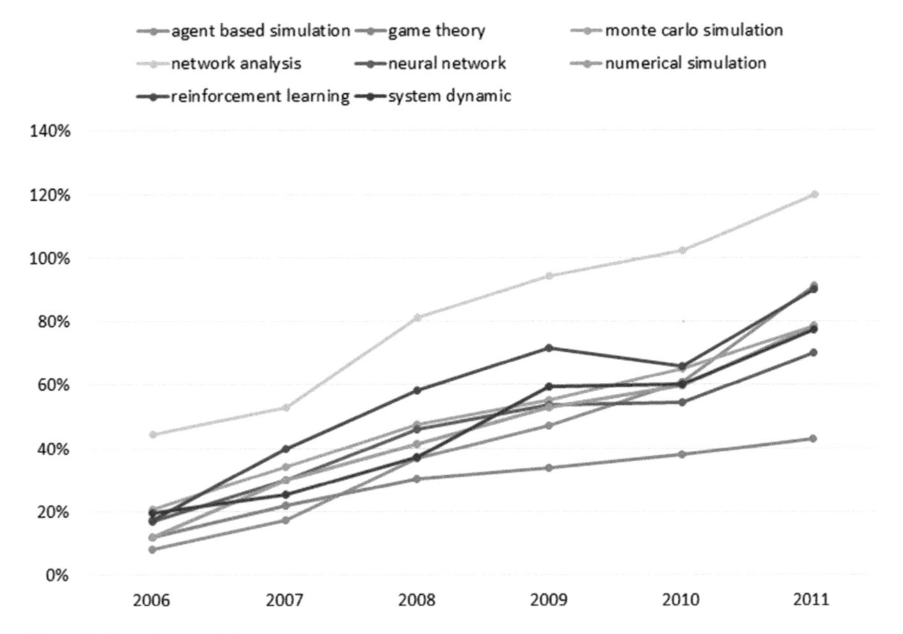

Fig. 8.5 Trend in statistics

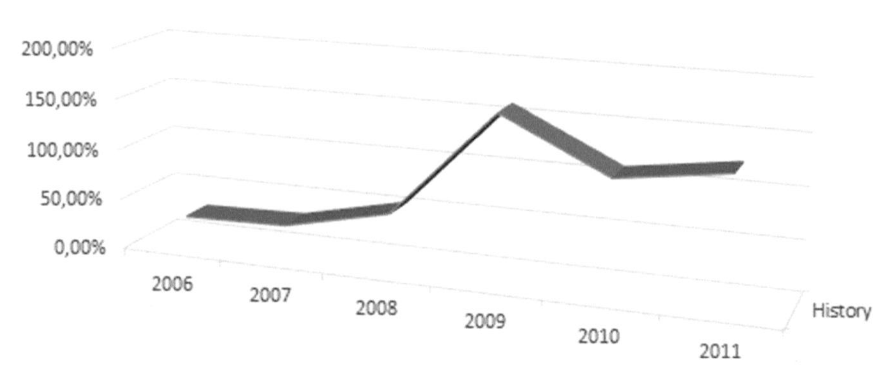

Fig. 8.6 History, using "agent based simulation" tag

knowledge, which study such matters,. Thanks to the study of both artificial and natural networks (the network of chemical reaction in a protein, the *web*, the network of citations among scientists), it was possible to highlight a new set of issues, in particular the concept of networks with topology different from the random topology, linking the idea of universality related to topological parameters (Albert and Barabási 2002; Barabasi et al. 1999, 2000; Strogatz 2001).

The situation is different for *agent-based simulation*, in which the increase does not depend on the universal character of the tag, but rather on a characteristic of the method. In fact, in our view, the use of models based on agents is increasing not because it yields new "universal" models, but rather because they these models to be *the only models that can work in a certain field of application* (Bankes 2002; Helbing 2012). Evidence in support of this argument lies in the value that the tag *agent-based simulation* adds to physics (see Fig. 8.4). Certainly, the interest of physics in the study of socio/economic phenomena (the so-called *econophysics* and *sociophysics* frameworks) has increased considerably in the last decade. Hence the necessity to find quantitative methods that can handle phenomena with highly heterogeneous agents; that are able to account for social influence; and that exhibit the ability to imitate, evolve, and evaluate different alternatives. ABM represents at least an attempt to find a solution to this problem (Bonabeau 2002).

The ascent of *reinforcement learning* and *Monte Carlo simulation*, on the other hand, is not clear: why do such *dated* tags persist over time? To clarify this apparent paradox, we should investigate possible correlations between them and a successful tag, like *agent-based simulation* and *reinforcement learning*. In fact, a main feature of agent-based modeling is the use of learning/adaptive agents (Andrighetto 2010; Campennì et al. 2009; Cecconi 2010; Epstein 2011).

History shows, for agent-based simulation, a typical pattern in these types of empirical studies: its sudden growth around 2009 could be explained as an artifact, a side effect of papers discussing the "*history of agent-based simulation*." In that period, indeed, we found a great deal of scientific production concerned with the "founding" problems of ABM. But it could be genuinely due to a historian's finding a new methodology (Dean 2000). The question is open for further investigation. One way to proceed might be to develop some algorithm to evaluate the text of the abstracts for the "top-ranked" papers, and to assign different weights to different semantic structures (for example, *history of agent-based simulation).*

In this paper we have shown how some *computational/simulation* tags, including agent-based simulation, occupy a larger space than other tags in the scientific literature, even in well- structured fields. This could indicate transversal trends in the growth of scientific paradigms.

The work is only beginning: it will be necessary to discover connections between scientific fields. For example, is it true that econophysics and sociophysics studies tend to abandon traditional methods of investigation—one-for-all statistical mechanics—in favor of simulations-based artificial agents? To give an answer it is probably necessary to try to understand the current trends in the scientific production and those that the scientific community will adopt in the near future both within the science of nature and that of society.

References

Albert, R., & Barabási, A.-L. (2002). Statistical mechanics of complex networks. *Reviews of Modern Physics, 74*(1), 47.

Andrighetto, G., Campenni, M., Cecconi, F., & Conte, R. (2010). *The complex loop of norm emergence: A simulation model 'Simulating interacting agents and social phenomena'* (pp. 19—35). Berlin: Springer.

Bankes, S. C. (2002). Agent-based modeling: A revolution?. *Proceedings of the National Academy of Sciences of the United States of America, 99*(Suppl 3), 7199–7200.

Barabási, A.-L., Albert, R., & Jeong, H. (1999). Mean-field theory for scale-free random networks. *Physica A: Statistical Mechanics and Its Applications, 272*(1), 173–187.

Barabási, A.-L., Albert, R., & Jeong, H. (2000). Scale-free characteristics of random networks: The topology of the world-wide web. *Physica A: Statistical Mechanics and Its Applications, 281*(1), 69–77.

Beel, J. (2009). Google Scholar's ranking algorithm: The impact of citation counts (An empirical study) 'Research Challenges in Information Science, 2009. RCIS 2009. International Conference on Information, 160–164, Technology: New Generations.

Bonabeau, E. (2002). Agent-based modeling: Methods and techniques for simulating human systems, *Proceedings of The National Academy of Sciences of the United States of America, 99*(Suppl 3), 7280–7287.

Campenni, M., Andrighetto, G., Cecconi, F., & Conte, R. (2009). Normal = Normative? The role of intelligent agents in norm innovation. *Mind & Society, 8*(2), 153–172.

Cecconi, F., Campenni, M., Andrighetto, G., & Conte, R. (2010). What do agent-based and equation-based modelling tell us about social conventions: The clash between ABM and EBM in a congestion game framework. *Journal of Artificial Societies and Social Simulation, 13*(1), 6.

Chen, X. (2010). Google scholar's dramatic coverage improvement five years after debut. *Serials Review, 36*(4), 221–226.

Conte, R. (2000). The necessity of intelligent agents in social simulation, *Advances in Complex Systems, 3*(01n04), 19–38.

Conte, R., Edmonds, B., Moss, S., & Sawyer, R. K. (2001). 'Sociology and social theory in agent based social simulation: A symposium'. *Computational & Mathematical Organization Theory, 7*(3), 183–205.

Dean, J. S., Gumerman, G. J., Epstein, J. M., Axtell, R. L., Swedlund, A. C., Parker, M. T., & McCarroll, S. (2000). Understanding Anasazi culture change through agent-based modeling, *Dynamics in human and primate societies: Agent-based* modeling of *social and spatial processes*, 179–205

Epstein, J. M. (2011). *Generative social science: Studies in agent-based computational modeling*. Princeton: Princeton University Press.

Helbing, D. (2012). *Agent-based modeling social self-organization.* Berlin: Springer, pp. 25–70.

Macy, M. W., & Willer, R. (2002). From factors to actors: Computational sociology and agent-based modeling. *Annual review of sociology,* 143–166.

Neuhaus, J. M., & McCulloch, C. E. (2006). Separating between- and within-Cluster covariate effects by using conditional and partitioning methods. *Journal of the Royal Statistical Society, Series B (Statistical Methodology), 68*(5), 859–872.

Squazzoni, F. (2008). The micro-macro link in social simulation. *Sociologica, 2*(1), 0–0.

Squazzoni, F. (2010). The impact of agent-based models in the social sciences after 15 years of incursions. *Sociological Methodology, 2*(2), 1–23.

Strogatz, S. H. (2001). Exploring complex networks. *Nature, 410*(6825), 268–276.

Chapter 9
On the Quality of Collective Decisions in Sociotechnical Systems: Transparency, Fairness, and Efficiency

Daniele Porello

9.1 Introduction

Decision-making in organization is a wide area that usually relies on formal methodology such as decision theory and game theory and on empirical investigations of actual decision-making in organizations. The aim of this paper is to propose a rather different question and to introduce a methodology to approach it: How can we conceptualize the quality of collective decisions made within the context of a complex sociotechnical system? Sociotechnical systems (STS) are complex organizational scenarios in which human agents interact in a normative constrained environment with themselves and with artificial agents (Emery and Trist 1960). For example, an understanding the organizational structure of an airport requires understanding the interaction between agents operating with metal detectors, sensors, and security cameras, as well as interacting with customers in a normatively specified way.

Defining STS is a complex task. Here we have decided to highlight the features of STS that are significant for understanding decision-making in this case. We view the complexity of STS as due to the entanglement of several layers of information—e.g., normative, perceptual, factual, conceptual—as well as of information sources, e.g., human, artificial, normative.

The quality of collective decisions in STS is evaluated by using the following three fundamental concepts: *transparency*, *fairness*, and *efficiency*. The key role of transparency in sociotechnical design was first stressed in (Guarino et al. 2012) and it has been argued that transparency is very important to enhance the adaptivity and resiliency of systems.

We conceptualize the *transparency* of a collective decision in terms of the entitlement of the agents involved in the systems to a justification of the decision made by that system. That is, the agents involved in the system (e.g., employee,

D. Porello (✉)
Laboratory for Applied Ontology, Institute of Cognitive Sciences and Technologies (ISTC-CNR), Via alla Cascata 56 C, 38123 Trento, Italy
e-mail: danieleporello@gmail.com

F. Cecconi (ed.), *New Frontiers in the Study of Social Phenomena*,
DOI 10.1007/978-3-319-23938-5_9

customers, users) are entitled to know the procedure that has been used to make the decision. Moreover, the choice of such a procedure has to be justified to them. Thus, a transparent decision has to be justified to those who are affected by the decision.

We conceptualize *justifications* of decisions in terms of *fairness* and *efficiency*. Intuitively, fairness is understood as non-arbitrary discrimination between the sources that are involved in the collective decision. For instance, a fair decision among stakeholders does not arbitrarily weight one's vote more than another. *Efficiency* is related to the rationality of the outcome. In decision theory or game theory, it is related to a maximization of an expected desirable value that is attached to the collective decision (Neumann and Morgenstern 1944).

We shall model fairness and efficiency conditions by means of techniques developed in welfare economics that have been recently used also in Multiagent Systems and Artificial Intelligence (Boella et al. 2011; Brandt et al. 2013; Woolridge 2008). In particular, we propose approaching the problem by using the methodology of *social-choice theory* (SCT) (Arrow 1963; Taylor 2005). SCT is a branch of welfare economics that studies the procedure for aggregating a number of possibly different individual preferences or choices into a collective preference or choice. An example of application of social-choice theory is voting theory, that is, the study of the property of voting procedures such as the majority rule. The reason that social-choice theory is a good methodology for investigating collective decisions is that it allows for specifying in a formal and clear way a number of properties that capture qualitative aspects of decisions. Those properties express, for instance, whether a procedure discriminates between individuals, whether the criterion of the choice has to be valid regardless the context of the decision, whether any issue to be decided has the same weight, and so on.

Moreover, social-choice theory provides an abstract treatment of collective decision-making that can be instantiated in a number of scenarios and allows us to check whether a certain procedure satisfies a number of qualitative desiderata. In particular, we shall use social-choice theory and judgment aggregation. The reason is that, as we shall see, those techniques provide versatile tools to model the aggregation of heterogenous types of information, and they allow for spelling out the properties of each type of aggregation procedure. The properties of aggregation procedure, or of decision procedures, then provide tools to model the concepts of justification of decisions that we look for.

Collective decisions are defined here not only as decisions made by a group or a team of individuals, such as committees, but also decisions that are made by the chief of a sector within the organization that is supposed to decide after gathering information coming from heterogeneous sources.

The application of social-choice theory to model collective decisions in sociotechnical systems requires a careful examination of the matter of possible decisions.

As we have recalled, a fundamental aspect of sociotechnical systems is the entanglement of heterogeneous layers of information. Therefore, we need to describe in an abstract and general way the types of information that are involved in complex sociotechnical systems.

In order to address and conceptualize this type of information, we shall use a foundational ontology. In particular, we shall exemplify our treatment by using DOLCE (Masolo et al. 2003, 2004) because it is capable of addressing the interconnection between different modules that gather different types of information, e.g., social, perceptual-mental, physical, organizational (cf. Boella et al. 2004; Bottazzi and Ferrario 2009; Porello et al 2014; Porello et al 2013).

The remainder of this paper is organized as follows. In Sect. 9.2, we informally discuss the background of social-choice theory and judgment aggregation. In Sect. 9.3, we present a model of judgment aggregation and we discuss the properties that formalize conceptions of fairness and efficiency. Section 9.4 presents our treatment of heterogeneous information in sociotechnical systems by means of DOLCE ontology. Section 9.5 approaches the problem of assessing the quality of decisions in sociotechnical systems by instantiating the methodology of judgment aggregation to possible scenarios of rich information entanglement.

9.2 Background on Social-Choice Theory and Judgment Aggregation

Social-choice theory originated through the seminal work of Kenneth Arrow (Arrow 1963), who provided a general framework for preference aggregation, namely, the problem of aggregating a number of individual conflicting preferences into a social or collective preference.

Take the following example: Suppose that a committee of three individuals (label them 1, 2, and 3) has to decide which security protocols to implement among three possible alternatives say: *a, b,* and *c*. In many settings of social-choice theory, preferences are assumed to be linear orders, that is, individual preferences are supposed to be *transitive* (an agent prefers *x* to *y* and *y* to *z*, then she/he should prefer *x* to *z*), *irreflexive* (an agent does not prefer *x* over *x*), or *complete* (for any pair of alternatives, agents know how to rank them, *x* is preferred to *y* or *y* is preferred to *x*).[1]

Suppose agents' possibly conflicting preferences can be faithfully represented by the following rankings of the options. Preference profiles are lists of the divergent points of view of the three individuals, as in the following example:

1. $a > b > c$
2. $b > a > c$
3. $a > c > b$

In the scenario above, the agents have conflicting preferences and there is no agreement on which is the best policy to be implemented. Since the policies are alterna-

[1] These conditions are to be taken in a normative way. They are not, of course, descriptively adequate, as several results in behavioral game theory show. However, the point of this approach is to show that even when individuals are fully rational—i.e., they conform to the rationality criteria that we have just introduced—the aggregation of their preferences is problematic.

tive, 1 and 3 would pursue a, whereas 2 would pursue b. In order to decide a collective option, we need a procedure that can settle the possible disagreement.

Suppose now that the individuals agree on a procedure to settle their differences; for example, they agree on voting by *majority* on pairs of options. Thus, agents elect the collective option by pairwise comparisons of alternatives. In our example, a over b gets two votes (by 1 and 3), b over c gets two votes (by 1 and 2), and a over c gets three votes. The majority rule defines then a social preference $a > b > c$ that can be ascribed to the group as the group preference.

The famous Condorcet paradox shows that it is not always the case that individual preferences can be aggregated into a collective preference. Take the following example:

1. $a > b > c$
2. $b > c > a$
3. $c > a > b$

Suppose agents again vote by majority on pairwise comparisons. In this case, a is preferred to b because of 1 and 3, b is preferred to c because of 1 and 2; thus, by transitivity, a has to be preferred to c. However, by majority also c is preferred to a. Thus, the social preference is not "rational," according to our definition of rationality, as it violates transitivity.

Kenneth Arrow's famous impossibility theorem states that Condorcet's paradoxes are not an unfortunate case of majority aggregation; rather they may occur for any aggregation procedure that respects some intuitive fairness constraint (Arrow 1963). In the next section, we shall discuss in more detail the formal treatment of the intuitions concerning fairness and we shall define a number of properties that provide normative desiderata for the aggregation procedure.

A recent branch of SCT, Judgment Aggregation (JA) (List and Pettit 2002; List and Puppe 2009) studies the aggregation of logically connected propositions provided by heterogeneous agents into collective information. The difference with preference aggregation is that in this case anti-type propositional attitudes can in principle be taken into account.

For example, take three sensors whose behavior can be described by the following propositions C "the alarm triggers" whenever A "metal is detected" or B "liquid is detected." In propositional logic this amounts to assuming that each sensor satisfies the constraint: $A \vee B \rightarrow C$

Suppose the three sensors 1, 2, and 3 provide different responses, each compatible with the above constraint.

	A	$A \vee B$	B	$A \vee B \rightarrow C$	C
1	Yes	Yes	Yes	Yes	Yes
2	No	No	No	Yes	No
3	No	Yes	Yes	Yes	Yes

In this case, a conflict may emerge from the fact that the three sensors may have divergent sensitivities on detecting A or B. One can study the aggregation procedure in order to define a notion of collective information provided by the aggregated behavior of the detectors.

In order to do that, one can choose a number of policies to aggregate sensors' information in order to define a sort of collective sensor. If we select *unanimity* in the example above, no proposition, besides the constraint, is elected as collective information, thus the collective sensor does not trigger any alarm. If the *majority rule* is used, then the collective information is given by all the propositions at issue; therefore the alarm triggers.

Analogously to the case of Condorcet's paradox in preference aggregation, situations of inconsistent aggregations of judgments have been individuated. These paradoxical situations have been labeled in the literature *doctrinal paradoxes* or *discursive dilemmas*. It is important to notice that such paradoxical situations actually occurred in the deliberative practice of the U.S. Supreme Court (Kornhauser and Sager 1993). This problem has been perceived as a serious threat to the legitimacy of group deliberation and it has been considered a seminal result in the recent debate on the rationality of democratic decisions (Kornhauser and Sager 1993; Pettit 2001).

We show an example of such a paradox by slightly modifying the previous example. Suppose 3 rejects B because she/he rejects the premise A.

	A	$A \vee B$	B	$A \vee B \rightarrow C$	C
1	No	No	No	Yes	No
2	Yes	Yes	No	Yes	Yes
3	No	Yes	Yes	Yes	Yes
Majority	*No*	*Yes*	*No*	*Yes*	*?*

By majority, A and B fail, so they are collectively false; however, the $A \vee B$ pass, which is inconsistent in classical logic. That would mean that the alarm triggers even in the case that none of A and B is collectively satisfied.

Such paradoxes does not exclusively concern the majority rule; they also apply to any aggregation procedure that respects some basic fairness desiderata. This is the meaning of the theorem proven by Christian List and Philip Pettit (List and Pettit 2002).

Therefore the notion of collective decision and collective information requires a careful examination of the aggregation procedures that provide viable solutions. In the next sections, we shall sketch a model for defining collective decisions, and we shall place it within sociotechnical systems.

9.3 A Model of Judgment Aggregation

We present the main elements of the formal approach of judgment aggregation (JA). The reason we focus on JA is twofold: on the one hand, it considered to be more general than preference aggregation (List and Pettit 2002); on the other hand, it has been claimed that JA can provide a general theory of aggregation of propositional attitudes (Dietrich and List 2009). Propositional attitudes, such as beliefs, desires, preference, and judgments, model the relationship between an agent and a sharable content.

Propositional attitudes have been extensively discussed in analytic philosophy, and formal languages for modeling propositional attitudes have been proposed by several contributions in philosophical logic (e.g., van Benthem 2011). Therefore, JA provides the proper level of abstraction for placing our model of decisions based on heterogeneous types of information.

Throughout this section, we shall refer to the individual sources of information in the system as individuals, who may represent actual human agent of the systems as well as sensors.

The content of this section is based on List and Pettit (2002) and Endriss et al. (2012) and builds on them. Let P be a set of propositional variables that represent the contents of the matter under discussion by a number of agents. The language L is the set of propositional formulas built from P by using the usual logical connectives (e.g. $\neg$, $\wedge$, $\vee$, $\rightarrow$).

Definition 1 An agenda X is a finite nonempty subset of L that is closed under (non-double) negations.

An agenda is the set of propositions that are evaluated by the agent in a given situation. In the examples of the previous section, the agenda is given by A, B, $A \vee B$, $A \vee B \rightarrow C$, C, plus their negations that allow us to model rejection of a certain statement: The rejection of a matter A is then modeled by an agent accepting $\neg A$. We define individual judgment sets as follows.

Definition 2 A judgment set J on an agenda X is a subset of the agenda J. We call a judgment set J **complete**, if for every formula in the agenda X, either A is in J or $\neg A$ is in J. We call J **consistent** if there exists an assignment that makes all formulas in J true.

We assume the notion of consistency that is familiar from logic. These constraints model a notion of rationality of individuals; i.e., individuals express judgment sets that are rational in the sense that they respect the rules of (classical) logic.

Denote with $J(X)$ the set of all complete consistent subsets of the agenda, namely, $J(X)$ denotes the set of all possible (rational) judgment sets on the agenda.

Given a set $N = \{1, \ldots, n\}$ of individuals, denote with $\boldsymbol{J} = (J_1, \ldots, J_n)$ a profile of judgment sets, one for each individual. A profile lists all the judgments of the agents who are involved in the collective decision at issue.

We can now introduce the concept of aggregation procedure. The domain of the aggregation procedure is given by $J(X)^n$, namely, the set of all possible profiles of individual judgments. The value of the aggregation function is assumed to be a set of judgment, i.e., an element of the power set $P(X)$.

Definition 3 An aggregation procedure for agenda X and a set of N individuals is a function F: $J(X)^n \rightarrow P(X)$.

An aggregation procedure maps any profile of individual judgment sets to a single collective judgment set. Given the definition of the domain of the aggregation procedure, the framework presupposes individual rationality: all individual judgment sets are complete and consistent. Note that we did not yet put any constraint on the collective judgment set, i.e., the result of aggregation, so that at this point the procedure may return an inconsistent set of judgments.

This is motivated by our intention to study both consistent and inconsistent collective outcomes. For example, in the doctrinal paradox of the previous section, the majority rule maps the profile of individual judgments into an inconsistent set. The consistency of the output of the aggregation is defined by the following properties.

Definition 4 An aggregation procedure F, defined on an agenda X, is said to be *collectively rational* if F is

- *complete* if $F(\boldsymbol{J})$ is complete for every profile $\boldsymbol{J}$ in $J(X)^n$;
- *consistent* if $F(\boldsymbol{J})$ is consistent for every profile $\boldsymbol{J}$ in $J(X)^n$.

That is, collective rationality forces the outcome of the procedure to be rational in the same sense of the individual rationality. Of course, the case of doctrinal paradox violates collective rationality.

We now introduce a number of properties— usually called axioms in social-choice theory—that provide a mathematical counterpart of our intuition on what a fair aggregation procedure is. The following are important axioms for JA discussed in the literature (Kornhauser and Sager 1993; List and Pettit 2002):

Unanimity (U): If for all agents i, a formula A is in J_i, then A is in $F(\boldsymbol{J})$.

Anonymity (A): For any profile $\boldsymbol{J}$ and any permutation of the individuals $\sigma: N \rightarrow N$, we have that $F(J_1, \ldots, J_n) = F(J_{\sigma(1)}, \ldots, J_{\sigma(n)})$.

Neutrality (N): For any formula A and B in the agenda and profile $\boldsymbol{J}$, if for all i we have that A is in J_i iff B is in J_i, then A is in $F(\boldsymbol{J})$ iff B is in $F(\boldsymbol{J})$.

Independence (I): For any formula A in the agenda and profiles $\boldsymbol{J}$ and $\boldsymbol{J}'$, if for all i, A is in J_i *iff* A is in J'_i, then A is in $F(\mathbf{J})$ iff A is in $F(\boldsymbol{J})$.

Monotonicity (M): If for any agent i, formula A in the agenda, and profiles $\boldsymbol{J}$ and $\boldsymbol{J'}$ such that coincide on every judgment set except for $J_{i,}$ we have that if A is not in J_i and A is in J'_i then if A is in $F(\boldsymbol{J})$, then $F(\boldsymbol{J'})$.

Such properties capture and formalize a number of intuitions concerning the fairness of the aggregation procedure. Unanimity entails that if all individuals accept a given judgment, then so should the collective. Anonymity states all individuals should be treated equally by the aggregation procedure. Neutrality is a symmetry requirement for propositions that prescribe that all the issues in the agenda have an equal weight. Independence says that if a proposition is accepted by the same subgroup under two distinct profiles, then that proposition should be accepted either under both profiles or under neither profile. Monotonicity entails that by adding support for a proposition, its acceptance does not change.

This fairness condition may be used to model the arguments that justify the collective decision to the individuals. For instance, it is well known by May's theorem (Taylor 2005) that the majority rule can be characterized in terms of those axioms:

the majority rule is the aggregation function that satisfies (A), (M), (N), plus a minimal rationality requirement (Endriss et al. 2012).

Therefore the justification of a decision made by majority may appeal to axioms such as (A), by saying that majority does not discriminate between individuals' opinions.

Of course there are situations in which the majority rule is not appropriate. For instance, when we know that the individuals providing information are not equally reliable, one may appeal to other axioms in order to justify the decision. A case for refraining from deciding by majority is when there are inconsistent outcomes. The methodology of judgment aggregation and social-choice theory allows us to know in advance what are the possible situations and the possible aggregation procedures that may lead to inconsistent outcomes. The impossibility theorem of List and Pettit (List and Pettit 2002) is as follows:

Theorem 1 (List and Pettit 2002) There are agendas such that there is no aggregation procedure that satisfies (A), (N), (I) and collective rationality.

In particular, for any aggregation procedure that satisfies (A) and (S), there is a profile of judgment sets that returns an inconsistent outcome. The majority rule that we have seen in the examples satisfies (A) and (N) and (I); accordingly, the discursive dilemma shows a case of inconsistent aggregation. Very simple agendas may trigger inconsistent outcomes, one example being the agenda of the doctrinal paradox that we have presented. Technically, any agenda that contains a minimal inconsistent set of cardinality greater than 2 may trigger a paradox.

A solution that would guarantee a rational outcome would be to use a dictatorship, i.e., a procedure such that a single individual in any possible scenario decides the outcomes. Such procedures are not desirable because, besides violating important intuitions concerning fairness, they amount to discharging all the relevant information of a given scenario.

The methodology of JA can be extended to treat many voting procedures and characterize whether they may return inconsistent outcomes. Moreover, since the notion of aggregation procedure is very abstract, one can in principle model more complex procedures or norms, such as those that define decision-making in organizations.

9.4 Ontological Analysis of Information in STS

A crucial aspect of decision-making in sociotechnical system is that decisions may concern and may be based on heterogeneous types of information. For instance, suppose a personnel director has to decide whether to fire an employee on the grounds that the employee is accused of theft. Further suppose that surveillance cameras seem to support the accusation, whereas human witnesses are against the accusation of theft. Moreover, such an accusation has a number of normative and procedural constraints that have to be satisfied in order to be effective. In such a case, a personnel director is faced with a decision that has to weight information coming from security cameras, human agents, and normative constraints, and then decide what to do.

In order to describe the complex layers of information that are possibly involved in sociotechnical systems, we need to integrate the perceptual, conceptual, factual, and procedural information into a harmonious system. We propose to use the DOLCE ontology as integrating framework (Masolo et al. 2003). After defining basic properties and relations that are generic enough to be common to all specific domains—like being an *object*, being an *event*, being a *quality*, or being an abstract (entity)—DOLCE specifies different modules, like the mental or the social module, that are composed of entities that share some characterizing features. For example, mental entities are characterized by being ascribable to intentional agents, and social entities are characterized by the dependence on collectives of agents. These conceptual relations specify the definitions of the basic entities in our ontology; e.g., roles are properties of a certain kind that are ascribable to objects (e.g., being employed by an organization).

In order to apply the ontology to a specific domain, we introduce domain-specific concepts that specify more general concepts belonging to all these modules (e.g., "an aircraft is a physical object").

The general ground ontology is meant to be not-context-sensitive and to provide a shared language to talk about some fundamental properties of concepts and entities. In this sense, the ontology provides a general language to exchange heterogeneous information and may be used as vocabulary to define communication languages for agents and to make explicit the matters of decisions.

We present some features of DOLCE-CORE, the ground ontology, in order to show that they allow for keeping track of the rich structure of information in a sociotechnical system.

The ontology partitions the objects of discourse, labeled particulars (PT) into the following six basic categories: objects, O; events, E; individual qualities, Q; regions, R; concepts, C; and arbitrary sums, AS. The six categories are to be considered rigid—i.e., a particular cannot change category through time. For example, an object cannot become an event.

In order to describe a concrete scenario for applying our ontological analysis, we enrich the language of DOLCE by introducing a specific language to talk about the scenario at issue. The language contains a set of *individual constants* for particular individuals. For example, in case we want to talk about an airport, individual constants may refer to "the gate 10," "the flight 799," "the landing of flight 747," or "the security officer at gate 10." Moreover, the language contains a set of *contextual predicates* that describe the pieces of information that agents may communicate in the intended situations (e.g., being a passenger, being a sensor, being a preference of an agent).

The language consisting of simple propositions can be partitioned according to the module they belong. For instance, we know that the predicates such as *passenger*, *customer*, *officer*, and *employee* can be accurately conceptualized as *roles*. Roles are social concepts that are characterized by the fact that they are anti-rigid (e.g., a passenger may cease to be a passenger) and dependent on other concepts (e.g. the concept of passenger requires the concept of person) (Masolo et al. 2004).

That is, in our specific ontology, we assume the axiom: RL (*employee*), that states that employee is a role. When we apply the predicate employee to an individual in our domain, e.g., *Employee* (Beatrix), we are building an atomic proposition that states some simple fact. This type of information can be retrieved by means of the ontological classification of the predicate. In this case, since employee is a role, it is a piece of social information belonging to the social module.

In a similar manner, we can list artificial sensors in our domain, e.g., *Sensor*(s1); categorize them as artificial agents, e.g., *ArtificialAgent*(s1); and model the output of a sensor as perceptual information coming from artificial agents.

We can easily extend the classification of predicates in order to partition all the (atomic) propositions into the relevant classes. For the sake of example, we can split here the possible types of propositions into *perceptual*, *social*, and *factual* propositions.

In Fig. 9.1, we depict a number of categories for an ontology developed in DOLCE for classifying information.

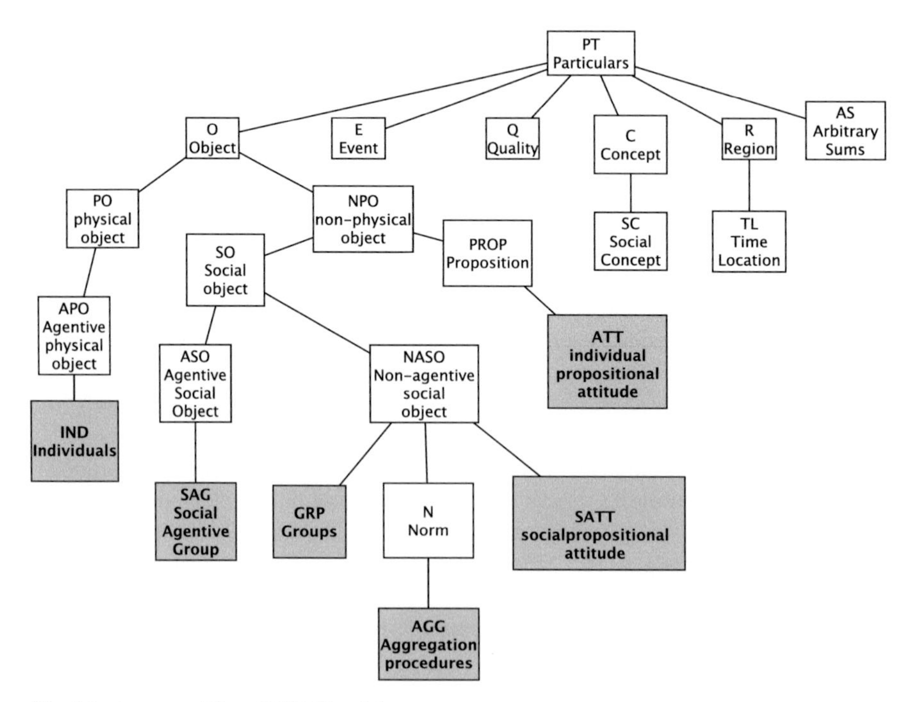

Fig. 9.1 An excerpt from DOLCE ontology

9.5 Assessing the Quality of Collective Decisions in Sociotechnical Systems

We have discussed how to represent in an abstract way the pieces of information that are required in order to provide an analysis of decisions and collective information in sociotechnical systems. We view agents as observation points in the system that are endowed with the reasoning capabilities provided by the ontology definitions in DOLCE and by logical reasoning. In this section, we present how to apply the methodology of JA to describe complex decisions in sociotechnical systems.

The properties of aggregation procedures that we have discussed in Sect. 9.3 provide a qualitative evaluation of the collective information or decision made within the system in a given moment. In a complex system like the one we are depicting, there may be several sources of disagreement between agents. For example, a possible disagreement may be at the level of perceptual information, as in the example of the sensors discussed in Sect. 9.2.

The ontological analysis allows us to classify the types of information; thus the question is how to evaluate the procedures that actually lead to collective decisions.

We briefly sketch our model. Suppose that we are able to list the agent—the information points—that are relevant for a certain decision. Call such a set of agents N of n agents. Denote as $A(L)$ the set of all possible sets of atomic formulas in our language L that are consistent with the ontology. We are presupposing that all the agents of the system agree on the definition provided by the ontological level. They may, however, disagree on matters of fact.

A profile of agents' propositional attitudes is given by a vector of sets of sentences, denoted $\boldsymbol{A}$. An aggregation procedure is a function F that takes a profile of agents' attitudes and returns a single set of propositions. The set of propositions $F(\mathbf{A})$ represents then the outcome of a collective decision of the system according to the procedure F.

For example, consider the case of the personnel director. Suppose there are three different security cameras and two human witnesses. Suppose proposition C means that "the accusation of theft is valid".

Agents	C
Camera 1	No
Camera 2	Yes
Camera 3	Yes
Human witness1	*Yes*
Human witness 2	*No*
Collective decision	*C in F(A)?*

Understanding what the procedure has been used to make the decision concerning C is crucial for the transparency of the system. We are not going to argue about which procedure is the best in this particular scenario. We claim only that social-choice theory and judgment aggregation, as well as the ontological analysis of information, allow for understanding and formalizing qualitative aspects of collective decisions in STS.

We now discuss a number of important concepts in evaluating collective decisions. In particular, we focus on the concept of *transparency*, the concept of *fairness,* and the concept of *efficiency* of decisions.

Firstly, a decision is *transparent* whenever the procedure F by means of which the decision has been taken is accessible to the agents involved in the decision.

In the example of the personnel director, the procedure is in fact *dictatorial*, because it is the director who has to take such a decision. However, what requires an explication, or even better, a justification, is the reason why the decision has been taken. That is, a dictatorial decision, such as the one taken by a single decision-maker, can nonetheless be a transparent decision, once it has been explained and justified to the relevant agents. One way of justifying such a decision is to mention how different information and different inputs affected the decision, which is equivalent to deciding which aggregation procedure a single decision-maker has followed with respect to different inputs. That is, in the example, the personnel director should make explicit whether the information coming from artificial agents outweighs the information coming from human agents.

The concept of *fairness* is quite debatable. However, the literature on social-choice theory is exactly about formalizing conceptions of fairness of an aggregation procedure. Therefore, the evaluation of fairness can be understood as the investigation of the properties of the decision procedures, for instance, whether the decision has been unanimous or anonymous with respect to the sources of information.

Unanimity implies that the agents of the system agree on a proposition. We claim that unanimity is a desirable property of any collective decision, regardless of the specific type of propositions. As agents are the observation points of the system, and our knowledge of the system is provided by means of agents' information, a violation of unanimity would amount to discharging information for no apparent reason (i.e., no agent against).

Anonymity, as we saw, implies that all agents are treated equally—we have no reason to weight the contribution coming from one agent more than the contribution coming from another one. This requirement is desirable when we cannot (or we do not want to) distinguish the reliability of agents. For example, we may not want to distinguish the information provided by two security officers that are communicating on the grounds of the higher reliability of the first compared to the reliability of the second. There are cases in which anonymity may not be a desirable property. For example, we want to weight the information coming from a trained security officer more than the information coming from a surveillance camera. Whenever appropriate, this is intended to model the fact that human agents may double-check outcomes from artificial agents, and human agents are assumed to be more reliable than artificial ones, at least at a number of tasks.

The condition of *independence* means that the acceptance of a formula at the systemic level only depends on the pattern of acceptance in the individuals' sets (e.g., the number of agents who accept). That is, the reason for accepting should be the same in any profile. Independence is a much more demanding axiom than the previous two; whether or not it should be imposed is debatable. A domain of application for which it is desirable is to merge normative information, where one expects impartiality across decisions.

Neutrality requires that all the propositions in the system have to be treated symmetrically. We believe that this is not desirable in the general case of heterogeneous information such as a STS. The reason is that we want in principle to treat visual, factual and conceptual information according to different criteria. Moreover, there are reasons to weight certain propositions more than others even when they belong to the same class. For example, the proposition that states that an object has been seen as a gun by a surveillance camera should be considered as highly sensible, and therefore it should be taken into account at systemic level. *Monotonicity* implies that agents' additional support for a proposition that is accepted at systemic level will never lead to it's being rejected. This property is desirable in most of the cases, provided the relevant agents are involved.

A further requirement that is usually viewed as a desirable property is the rationality of the collective decision. In particular, we focus on consistency: An aggregator F is consistent if for every profile, the set $F(A)$ is consistent with the ontology. As we saw, not every aggregator that satisfies the properties that we have seen guarantees consistency. For example, merging information by means of the majority rule or by a quota rule may lead to inconsistent sets of propositions.

The concept of consistency models a very weak notion of *efficiency* and more demanding views on efficient decisions can be modeled by adding further constraints.

We conclude by presenting a class of procedures that can be tailored for aggregating information in the scenario of STS. Those procedures are discussed in detail in (Porello and Endriss 2014) and (Taylor 2005).

Given a set of propositions X, we define a priority order on formulas in X as a strict linear order on X. Several priority orders can be defined on X, for example, a *support* order ranks the propositions according to the number of agents supporting them. Moreover, a *relevance* order ranks types of propositions (e.g., factual, perceptual, normative) according to their importance for the decision at issue. Moreover, we can define a priority order on propositions that depends on the *reliability* of the agents that support them. Thus, the reliability priority may be defined as a proposition A is more reliable than B if the number of experts supporting A is greater than the number of experts supporting B.

Thus, a priority-based procedure tries to provide a consistent outcome by checking the relevant information according to the priority. That is, the procedure tries to discharge conflicting information with a lower priority. For priority-based procedures, neutrality or anonymity may be violated by the priority order. Independence is also violated (because it may cease to be accepted if a formula it is contradicting receives additional support). Moreover, such procedures ensure consistency by construction.

Priority-based procedures allow for weighting the information according to the reliability or the relevance of different sources. For example, we can weight the information coming from security officers, who are viewed as experts, more than information coming from surveillance cameras. Moreover, we can weight the reports of cameras that are closer to the location at issue more than the information coming from other cameras. Note that it may be hard to compute the systemic information, given the required consistency check. The complexity depends of course on the language that we use to implement our ontology (a study of the complexity of computing problems related to judgment aggregation was presented in Endriss et al. 2012).

It is interesting to point out an application of non-consistent aggregators, namely, aggregators that return inconsistent sets of propositions. By using the analysis of aggregators provided by judgment aggregation, it is possible to pinpoint the places where the inconsistencies in the system are generated. In particular, aggregators that may return inconsistent information are useful to pinpoint causes of normative or conceptual disagreement, namely, to analyze incompatibility of norms or concepts defined in the system with the collective information gathered by the agents.

9.6 Conclusions

We have presented some basic elements for developing a model for assessing the quality of collective decisions in sociotechnical systems. We argued that we need a precise ontological understanding of the pieces of information involved in decisions and that welfare economics, social-choice theory, and judgment aggregation provide important tools for understanding fairness and efficiency of decisions. Therefore, foundational ontology plus the study of aggregation procedures provide important elements for developing a theory of justification of collective decision.

As a conclusion, we can view transparency as a necessary condition in order to make an assessment of the quality of decisions possible. Transparency amounts to making the procedure and the motivation of a collective decision accessible. That is, the first thing we need to demand in a system is transparency. We conceptualized transparency as a form of entitlement of the agents involved in the system to a justification of the decision made by the system. Future work has to investigate this concept in detail. For instance, one further condition on justifications is that they have to be addressed to real agents; that is, they have to be accessible to them—for instance, they have to be cognitively adequate to their addressees. Moreover, justifications have to be acknowledgeable by real agents; they should appeal to reasons that are shared among agents.

Acknowledgments D. Porello is supported by the VisCoSo project, financed by the Autonomous Province of Trento, "Team 2011" funding program.

References

Arrow, K. (1963). Social choice and individual values. Cowles foundation for research in economics at Yale University, Monograph 12. Yale: Yale University Press.

Boella, G., Lesmo, L., & Damiano, R. (2004). On the ontological status of plans and norms. *Artificial Intelligence and Law, 12*(4), 317–357.

Boella, G., Pigozzi, G., Slavkovik, M., & van der Torre, L. (2011). Group intention is social choice with commitment. In Proceedings of the 6th international conference on coordination, organizations, institutions, and norms in agent systems, COIN@AAMAS'10, pp. 152–171, Berlin, Heidelberg, Springer-Verlag.

Bottazzi, E., & Ferrario, R. (2009). Preliminaries to a DOLCE ontology of organizations. *International Journal of Business Process Integration and Management, Special Issue on Vocabularies, Ontologies and Business Rules for Enterprise Modeling, 4*(4), 225–238.

Brandt, F., Conitzer, V., & Endriss, U. (2013). Computational social choice. In G. Weiss (Ed.), *Multiagent systems*. Cambridge: MIT Press.

Dietrich, F., & List, C. (2009). The aggregation of propositional attitudes: Towards a general theory. Technical report.

Emery, F. E., & Trist, E. L. (1960). Socio-technical Systems. In C. W. Churchman & M. Verhulst (Eds.), *Management science, models and techniques* (Vol. 2, pp. 83–97). Pergamon.

Endriss, U., Grandi, U., & Porello, D. (2012). Complexity of judgment aggregation. *Journal of Artificial Intelligence Research, 45,* 481–514.

Guarino, N., Ferrario, R., & Sartor, G. (2012). Open ontology-driven sociotechnical systems: Transparency as a key for business resiliency. In M. De Marco, D. Te'eni, V. Albano, & S. Za (Eds.), *Information systems: Crossroads for organization, management, accounting and engineering*. Berlin: Springer.

Kornhauser, L. A., & Sager, L. G. (1993). The one and the many: Adjudication in collegial courts. *California Law Review, 81*(1), 1–59.

List, C., & Pettit, P. (2002). Aggregating sets of judgments: An impossibility result. *Economics and Philosophy, 18*, 89–110.

List, C., & Puppe, C. (2009). Judgment aggregation: A survey. In P. Anand, C. Puppe, & P. Pattanaik (Eds.), *Handbook of rational and social choice*. Oxford: Oxford University Press.

Masolo, C., Borgo, S., Gangemi, A., Guarino, N., & Oltramari, A. (2003). Wonderweb deliverable d18. Technical report, CNR.

Masolo, C., Vieu, L., Bottazzi, E., Catenacci, C., Ferrario, R., Gangemi, A., & Guarino, N. (2004). Social roles and their descriptions. In Proc. of the 6th Int. Conf. on the principles of knowledge representation and reasoning (KR-2004), pp. 267–277.

Neumann, J. V., & Morgenstern, O. (1944). *Theory of games and economic behavior*. Princeton: Princeton University Press.

Pettit, P. (2001). Deliberative democracy and the discursive dilemma. *Philosophical Issues, 11*(1), 268–299.

Porello, D., & Endriss, U. (2014). Ontology merging as social choice: Judgment aggregation under the open world assumption. *Journal of Logic and Computation, 24*(6), 1229–1249.

Porello, D., Setti, F., Ferrario, R., Cristani, M. (2013). Multiagent socio-technical systems: An ontological approach. In Proceedings of COIN@AAMAS/PRIMA 2013, pp. 42–62

Taylor, A. D. (2005). *Social choice and the mathematics of manipulation*. New York: Cambridge University Press.

van Benthem, J. (2011). *Logical dynamics of information and interaction*. Cambridge: Cambridge University Press.

Woolridge, M. (2008). *Introduction to multiagent systems*. New York: Wiley.

Chapter 10
How Crime Spreads Through Imitation in Social Networks: A Simulation Model

Valentina Punzo

10.1 Introduction

Social influence has been assumed to play an important role in the explanation of crime (for example, Sutherland 1947; Sutherland and Cressey 1966; Burgess and Akers 1966). Mechanisms of social influence explain *how* the social environment affects individual crime decisions.

The criminology/sociology literature has traditionally relied on the immediate micro-level social environment of the individual to explain his or her behavior (see Sutherland 1947). The main argument is that criminal and deviant behavior is learned in interaction with others. Although a number of studies have reported empirical evidence about the effects of social influence on the levels of criminal offending as well as on the spatial variance of crime rates (Glaeser et al. 1996; Sellers et al. 2003; Brezina and Piquero 2003; Chappell and Piquero 2004; Triplett and Payne 2004), some questions still arise about the *process* by which social learning takes place.

Recently, several social theorists have paid specific attention to *imitation* as the most important behavioral process by which the learning of criminal behavior comes about within social networks (Burgess and Akers 1966; Akers 1985, 1998; Akers and Jensen 2006; Ormerod and Wiltshire 2009).

Our study aims to shed light on the *mechanisms* of imitation and on their effects on crime. In other words, the purpose of the paper is to examine the extent to which the emergence of crime can be explained as a social-network phenomenon.

We investigated two different mechanisms of imitation: *rational imitation* and *social imitation*. In order to test our hypothesis we used an agent-based approach. In the model individual agents interact in their social networks and their decisions to be engaged in crime, including their consequent behavior towards crime, are

V. Punzo (✉)
Department of Law, Society and Sport, University of Palermo,
Via Maqueda 172, 90134 Palermo, Italy
e-mail: valentinapunzo@libero.it

F. Cecconi (ed.), *New Frontiers in the Study of Social Phenomena*,
DOI 10.1007/978-3-319-23938-5_10

influenced by both personal and social learning factors. The simulation investigates whether there are any conditions in which these mechanisms of imitation, also in relation with *social network topologies*, could affect individual criminal choices.

Before presenting the agent-based model, the theoretical framework on which our simulation relies is introduced.

10.2 Literature Review

Our current understanding of the role of the social environment in crime causation is undeveloped (Sampson et al. 1997; Sampson and Wikström 2008). This is partly a consequence of the lack of well-developed theoretical models for *how* social environments influence people's engagement in acts of crime (Wikström et al. 2010, p. 56).

The role of the social environment (Cohen and Felson 1979; Felson 2002; Wikström 2006; Wikström and Treiber 2007, 2009) is crucial within the explanatory framework of situational models of crime. For example, *Routine Activity Theory* (Cohen and Felson 1979; Felson 2002) and *Situational Action Theory* (Wikström 2006, 2010) suggest that some social environmental conditions are more *criminogenic* than others, representing *opportunities for crime* (Cohen and Felson 1979; Clarke 1997) and make specific predictions for *how* the interaction between a person's propensity and social environmental exposure causes acts of crime (Wikström and Treiber 2009; Wikström et al. 2010). According to these approaches, acts of crime are an outcome of the convergence between people and setting (Wikström et al. 2010).

According to *Situational Action Theory*, "the likelihood an act of crime will be committed by a particular person in a particular setting depends upon the extent to which that person's moral rules and the moral rules of that setting are consistent with the rules of conduct defined by law" (Wikström et al. 2010, p. 61)[1]. Individuals often look to social norms both to gain an accurate understanding of and effectively respond to social situations and to create and maintain meaningful social relationships with others (the so called *affiliation-oriented goal*) (Cialdini and Goldstein 2004).

Thus, the social norms of the social groups in which a person takes part and their enforcement (through the process of deterrence) are the causally relevant social environmental features that determine *criminogenic exposure*, that is, a moral context conducive to crime (Wikström 2004).[2] Accordingly, the concept of *collective*

[1] *Setting* is a key concept of the theory. It refers to the part of the environment which an individual can, at a particular moment in time, access with his or her senses (Wikström 2004).

[2] In a recent empirical study on the role of the social environment in crime causation, Wikström et al. (2010) found that young people with higher crime propensity (based on a crime-prone morality and low ability to exercise self-control) are more frequently exposed to criminogenic settings (which are encountered more often by young people when spending time in settings outside the home and school areas). Those who spend more time in criminogenic settings (e.g., being unsupervised with peers in areas with a poor collective efficacy) tend to be more frequently involved in acts of crime. However, and importantly, this relationship depends on the young person's crime propensity (Wikström et al. 2010).

efficacy (Sampson et al. 1997, 1998), advanced by modern social disorganization theorists, measures a key aspect of the moral context (the level of enforcement of relevant moral rules), usefully implemented for the explanation of the neighborhood effects on crime (Sampson and Wikström 2008; see Elliott et al. 1996; Sampson et al. 1998).

The invocation of the immediate micro-level social environment of the individual to explain his or her behavior is crucial within the explanatory framework of the *theory of differential association* (Sutherland 1947), which stresses the social influence processes that underlie criminal activities and advocates the idea that the impact of social norms on decision processes is influenced by *learning processes*. Criminal behavior patterns are thus *learned* in interaction with others who are deviant.

In other words, *Differential association theory* can be seen as a specific instance of the more general network theory of social learning, that an individual's attitudes and behavior are affected by the attitudes and behaviors of the members of his or her personal network, and the effects are conditioned by the characteristics of the network. Specifically, social learning theory points out that "important" or *prestigious* contacts have a larger influence on learning criminal behavior.

The role of social influence processes as well as the impact of differential associations and social networks have been highlighted by empirical studies on a range of minor deviance, substance use, delinquent behavior, and serious crimes (Katz et al. 2001; Ludwig et al. 2001; Akers and Jensen 2006; Akers and Sellers 2009; for review see Akers and Jennings 2009; Glaeser et al. 1996, 2008; Haynie 2002; Warr 2002).[3]

Basically, the general conclusions taken from all these studies are that criminal behavior is affected not only by individual incentives but also by actions performed by others (i.e. peers, neighbors), the so called "reference group" (Scheinkman 2008).

In other words, social networks are a natural way to explain the emergence of deviance as well as the levels of criminal offending (Calvo-Armengol and Zenou 2004; Bruinsma and Bernasco 2004; Ormerod and Wiltshire 2009).

Some extensions of the early social learning approach focused on *imitation* as the most important behavioral process by which the learning of criminal behavior takes place (Burgess and Akers 1966; Akers 1985, 1998; Akers and Jensen 2006).

As a specific mechanism of social learning, where agents learn by observing choices made by other agents (Scheinkman 2008), imitation may be then presumed to require *copying* at least (Hurley and Charter 2005). According to Hedström, "an actor A can be said to *imitate* the behavior of actor, B, when the observation of the behavior of B affects A in such way that A's subsequent behavior becomes more similar to the observed behavior of B" (Flanders 1968 cit. in Hedström 1998, p. 307). Following Hurley and Charter, "the observers' perception of the model's behavior causes similar behavior in the observer, in a way such that the similarity between the model's behavior and that of the observer plays a role, though not necessarily at a conscious level, in generating the observer's behavior" (Hurley and Charter 2005, p. 2).

[3] In their cross-sectional model, Glaeser et al. (1996) have shown that more than 70% of the spatial variation of crime against property (both inter- and intra-city) can be explained by social interactions instead of differences in local attributes.

Among the main perspectives on imitative behavior, some of them have pointed out the role of social capabilities, such as social experience, in the development of the capacity for imitation (see Heyes 1999; Heyes and Ray 2000). Others approaches, on the contrary, have underlined cognitive capabilities in social learning (Bandura 1977).

From a sociological point of view, social-relationship-oriented motivations are relevant in the explanation of imitative behavior. In this perspective, Cialdini highlighted the role of the so-called *conformity motivations*-based on the goal of obtaining social approval from others, to build rewarding relationships and maintaining one's self-concept (2001; Cialdini and Goldstein 2004). Thus, the motivation to affiliate with others affects the extent to which a certain behavior is imitated (Lakin and Chartrand 2003; Chartrand and Bargh 1999). In other words, imitation relies on *social power*, where "individuals are frequently rewarded for behaving in accordance with the opinions, advice, and directives of authority figures" (Cialdini 2004, p. 595).

Research on social interactions reveals several problems in the conceptualization of *imitation*. As suggested by Manski (1993, 2000), it is not at all obvious to empirically demonstrate that peer interaction is responsible for the positive statistical association observed between the behavior of an actor A and that of an actor B (see also Manzo 2013). In fact, persons in the same group tend to behave similarly because they share some similar individual characteristics (the so called "correlated effect," see Manski 1993, p. 31), or they are exposed to similar exogenous stimuli, such as social background characteristics, social environments (see the concept of "contextual interactions": Manski 2000, p. 23). Following this approach, in his computational study on educational choice, Manzo (2013, p. 51) hypothesized that a "similar (educational) outcome may arise not from the influence that the two actors exert on each other, which would constitute the interaction-based 'endogenous effect' in which one is interested, but from the potentially unmeasured shared factors that modify the probability of being friends and that of experiencing a certain (educational) outcome."

Despite growing attention on imitation, the *causal mechanisms* that link imitative processes to crime decisions are still poorly understood (Laland and Bateson 2001).

The present study aims to explore the imitative learning mechanisms involved in social interactions in order to study their effects on individual deviant/criminal choices and on the patterns of the spreading of crime. It is then possible to recognize different *mechanisms* (Hedström and Swedberg 1998; Hedström 2005; Manzo 2007) of imitative behavior, where not all behavioral patterns are equally imitated.

We hypothesized different criminal outcomes, at a macro-level, generated by different learning mechanisms of imitation involved at the micro-level of social interaction. Specifically, we investigated two different mechanisms of imitation: *rational imitation* (Hedström 1998; Schwier et al. 2004) and *social imitation*.

In order to investigate our hypothesis we used an agent-based computational approach. In our view, *rational imitation* refers to a situation "where an actor acts rationally on the basis of beliefs that have been influenced by observing the past choices of others. To the extent that other actors act reasonably and avoid alternatives that have proven to be inferior, the actor can arrive at better decisions than he or she would make otherwise by imitating the behavior of others" (Hedström 1998, p. 307). Several studies on early imitation found that the intentional structure underlying the imitative learning behavior reproduces the result that the modeled actor intended to achieve (Meltzoff 1995). This finding is most often interpreted as revealing that *rational imitation* is a genuine imitative learning with a flexible intentional structure relating observed means to observed results. The capacity to *copy* observed results may underlie an early understanding of action in terms of goals and intentions (Hurley and Charter 2005, p. 32).

Accordingly, in a sociological perspective on crime, antisocial models are easily *mimicked*, according to the *reinforcement principles* of learning theory (Akers 1966, 1985, 1998). In this framework, the likelihood that an individual will imitate an observed behavior is contingent on any observed *consequences* that resulted from the model's behavior (Akers and Jennings 2009, p. 109). Thus, a certain criminal behavior is imitated because past examples have been rewarded (Burgess and Akers 1966).

Consistent with this view, in our simulation, *rational imitation* is based on the *performance* observed.

Conversely, *social imitation* refers to a situation in which an actor imitates the behavior of actors *highly integrated* into the network. Starting from Barabási and Albert's (1999) work on complex network growth, several studies have showed that most real social networks present a common structure: few nodes (network elements) are highly connected into the network and many nodes are poorly connected. The *preferential attachment* mechanism (Barabási and Albert 1999) at the basis of growing social networks means that the higher the degree of a node, the more new edges the node will attract (Lowe 2009). This model is also consistent with the social learning approach to deviance, for which actors highly integrated into a network (the so called *hubs*) perform the function of socialization to deviance (Becker 1963, 1967).

Social imitation refers to those *prestigious* contacts (measured in terms of contacts or links with peers) that, according to social influence approach, have the largest influence on learning criminal behavior (Akers 1985).

In our simulation environment, *social imitation* is then based on *social prestige* acquired by those who are strongly socially embedded into the network.

Employing a computational model, we directly observed the different social outcomes, in terms of criminal behavior, generated by both mechanisms, i.e., *rational imitation* and *social imitation*.

10.3 The Agent-Based Approach

Agent-based social simulation (ABSS) has increasingly proved to be successful for the study of crime and deviance (Liu et al. 2005; Wang 2005; Liu and Eck 2008; Birks et al. 2008). The main purpose of agent-based modeling is to analyze the properties of social systems by explicitly representing individuals (called agents) and the interactions between them and the (geographical, spatial, economic, institutional) environment in which they are situated (Miller and Page 2007; Squazzoni 2008; Gilbert 2008).

In crime modeling, *agents* represent criminals (or potential criminals), potential victims, police, and/or other informal control agents. Agents make decisions about movement and actions in a local *environment* (for example, a street network and/or a social network). It is then possible, as we did in our model, to simulate social interactions between different decision-makers embedded in social networks and to observe the emergence of macro-level crime patterns (Groff 2007).

For the purpose of our study, agent-based models (ABM) seemed to us a suitable method for two main reasons. First of all, agent-based simulations can be usefully employed to investigate the *mechanisms* that give rise to a certain social phenomenon, as, for example, crime spreading, rather than exactly reproduce it (Sawyer 2003; Manzo 2004, 2007; Hedström and Åberg 2006). In fact, crime simulations allow researchers to examine not only the mere distribution of crime patterns but also *how* they develop (for example, those mechanisms that give rise to crime patterns or prevent crime from clustering). A few, simple, theory-based rules that inform the behavior of individual agents (and their interactions) *generate* macro-level patterns (Gilbert and Troizsch 2005).

Secondly, simulated experiments through ABM help criminologists to face the weakness of theoretical explanations of crime because they provide a rigorous formalization of a certain theory and explanation, useful for experimentation. Thus, experimenting with artificial crime models may help to formulate hypotheses about *how* crime is produced (Wang et al. 2008; Groff 2008a).

In this perspective, Bosse and colleagues have implemented an agent-based model to simulate the process of social learning of deviance and to test some assumptions of *differential association theory* (Bosse et al. 2009). Moreover, some recent models have included resources made available in the area of social network analysis, characterizing the impact of social network topologies (i.e., scale-free networks, small-world settings) on the development and growth of special types of crimes (Kaza 2005; Furtado et al. 2008; Ormerod and Wiltshire 2009).

Following this approach, agent-based modeling allows us to observe in a more formal and analytical way the structure of social networks and to investigate *how* imitation mechanisms come about on social networks as well as *how* they affect the spread of crime on social networks. ABM are then used to investigate the conditions to account for the spreading of crime.

In our simulation, we modeled individual agents who face different criminal/deviant opportunities (i.e., gambling, heavy drinking, drug use, shoplifting, etc.). Individuals interact in their social networks, influencing each other by imitation. Agent behavior is influenced by both individual and social learning factors.

We investigated the effects of two different mechanisms of imitation, *rational imitation* and *social imitation*, both on individual criminal choices and on the pattern of the spreading of crime. Through controlled simulated experiments we could indeed observe differences in the behavior toward crime emerging as the result of both rational imitation and social imitation especially in relation with network topologies.

10.4 The Simulation Model

Our simulation provides an agent-based model, implemented using the NetLogo simulation environment (Wilensky 1999), in which we modeled the person–environment interaction, starting from the assumption that acts of crime are an outcome of the convergence between people and setting (Wikström et al. 2010).

The model structure includes an environment populated by a limited number of objects that represent *opportunities* for crime (or also called *criminal opportunities).* Numerical entities (hereafter called "artificial agents" or, simply, "agents") were programmed to move around the environment and to choose whether to take criminal opportunities they encountered. In the present thematic context, each choice represents a decision about whether to engage in criminal behavior or not. Agents can only make a criminal choice if they encounter an opportunity to commit a crime. Moreover an agent's decision is based on personal and social learning factors.

The Netlogo model includes two kind of objects: *criminal opportunities* displaced in the simulation environment and *individual agents* who move around. The artificial agents are assumed to mimic the real actors. Then they are exogenously attributed to some social networks and they can imitate the behavior of other agents present within their network.

In fact, individuals interact within social networks, influencing each other through the mechanisms of imitation. In the model, aggregate deviant dynamics observed emerge from individual deviant choices which *evolve* through the social learning mechanisms of imitation.

Criminal opportunities[4] are characterized by different combinations of costs, benefits, and probability of success: some offer high *benefits* and/or low *costs* (and are therefore more attractive); some offer high *costs* and/or low *benefits* (and are therefore less pleasurable); some offer high risks (some opportunities are therefore associated with a lower *probability of success*)[5].

The properties of criminal opportunities, continuous variables uniformly distributed, follow:

---

[4] The situational model of *Routine Activity Theory* (Cohen and Felson 1979; Felson 2002) defines what constitutes an *opportunity* as the convergence of a motivated offender and a suitable target in the absence of guardianship (supervision, control) (Cohen and Felson 1979). Looking at crime as a situated event, situational models of crime are based on the premise that some situations are more favorable for crime than others (see Birbeck and La Free 1993). Accordingly, *criminal opportunity* in the model refers to the crime event, that is the situation that can be more or less favorable, attractive or advantageous.

[5] Although the economic approach to crime typically focuses on economic outcomes (Becker 1968), in a broader sociological rational choice perspective (Cornish and Clarke 1986, 1987, see McCarthy 2002), *benefits* and *costs* of criminal opportunities are interpreted not only in economic or legal terms (i.e., crime's financial returns, illegal incomes, or economic/punishment costs, such as arrests) but they also include several social factors such as excitement, on the one hand, and non-legal sanctions related to social reputation or "moral" costs related to individual conscience and interiorized norms on the other (Grasmick and Bursik 1990; Mc Carty 2002; Mehlkop and Graeff 2010). In this regard, there is growing agreement about the relevance of social costs, typically more important than those associated with imprisonment and loss of wages (Nagin 1998).

Benefits: [0,1] incentives associated with criminal opportunities. The benefits value indicates the amount of individual payoff that is increased when an actor takes a criminal opportunity and wins;

Costs: [0,1] legal and non-legal costs associated with each criminal opportunity. The costs value indicates the amount of individual payoff that is decreased when an agent takes a criminal opportunity and loses;

Probability of success: [0,1] the probability of carrying out the offense successfully, that is, the probability of not being convicted. It refers to the probability of winning that is associated with each criminal opportunity.

An agents' decision to take a criminal opportunity is influenced by some estimates of a criminal opportunity's costs and benefits,[6] on the basis of their attitudes toward risk, their desire to achieve a goal and subjective expectations.[7] Moreover, an actor's intentional actions are guided by his attitudes toward social norms (moral values). Thus, actors choose a certain action if they positively evaluate it and if they expect their peers to advocate this behavior (Wikström 2006).[8] This means that the agents' decisions about crime are somewhat positively correlated.

Following these propositions, in the model, benefits and costs perception will be affected by a *bias* linked to the agent. Thus every agent in the model decides whether to undertake a deviant action by performing an evaluation of costs and benefits of the opportunity for crime, on the basis of their *bias*.

Agents' properties are:

Bias [−1, 1] (continuous variable uniformly distributed) is the individual attitude to perceive the costs and benefits that are associated with criminal opportunities (some with a low bias overestimate the cost and underestimate the benefit; some with a high bias overestimate the benefit and underestimate the costs);

Action $\{-1, 0, 1\}$ is the outcome of individual choice. It is a property that defines whether the agent has taken a criminal opportunity. It can be worth -1 if the agent decided not to act; 1 if the agent decided to act; or 0 if the agent had no opportunity to commit a crime;

Payoff is the score amount of each agent (see Fig 10.1, Simulation process diagram, behavior phase).

---

[6] Rational-choice theory provides a fruitful approach to understanding criminal decision-making. According to this theory, individuals commit crimes because of their different costs and benefits. Thus, the choice of a criminal action is determined exactly by the varying assessments of costs, risks, and utility by different potential offenders (Cornish and Clarke 1987). Firstly, an actor chooses to undertake a deviant/criminal action when they *subjectively expect* it will increase their benefit (*Ibid*., p. 933); secondly, benefits and costs of criminal opportunities that are identical (or we can say objective) can be evaluated differently by different actors (*Ibid*., p. 935). A prison sentence, for example, might be subjectively experienced differently by different people, where the decision to commit a crime is affected by subjective perceptions or assessments. Then, subjective assessments will be objectively accurate only within the limits of individuals' bounded rationality (see Cornish and Clarke 1987; Simon 1993).

[7] Experiences or *differential associations* (Sutherland and Cressey 1966) also contribute to the formation of subjective expectation. Association with successful bank robbers, for example, would encourage actors to assume there is a low probability of being convicted (Mehlkop and Graeff 2010).

[8] According to Wikström (2006), *criminogenic settings* present a moral context conducive to crime, which influences a person's perception and the consequent choice of criminal action (Wikström 2004).

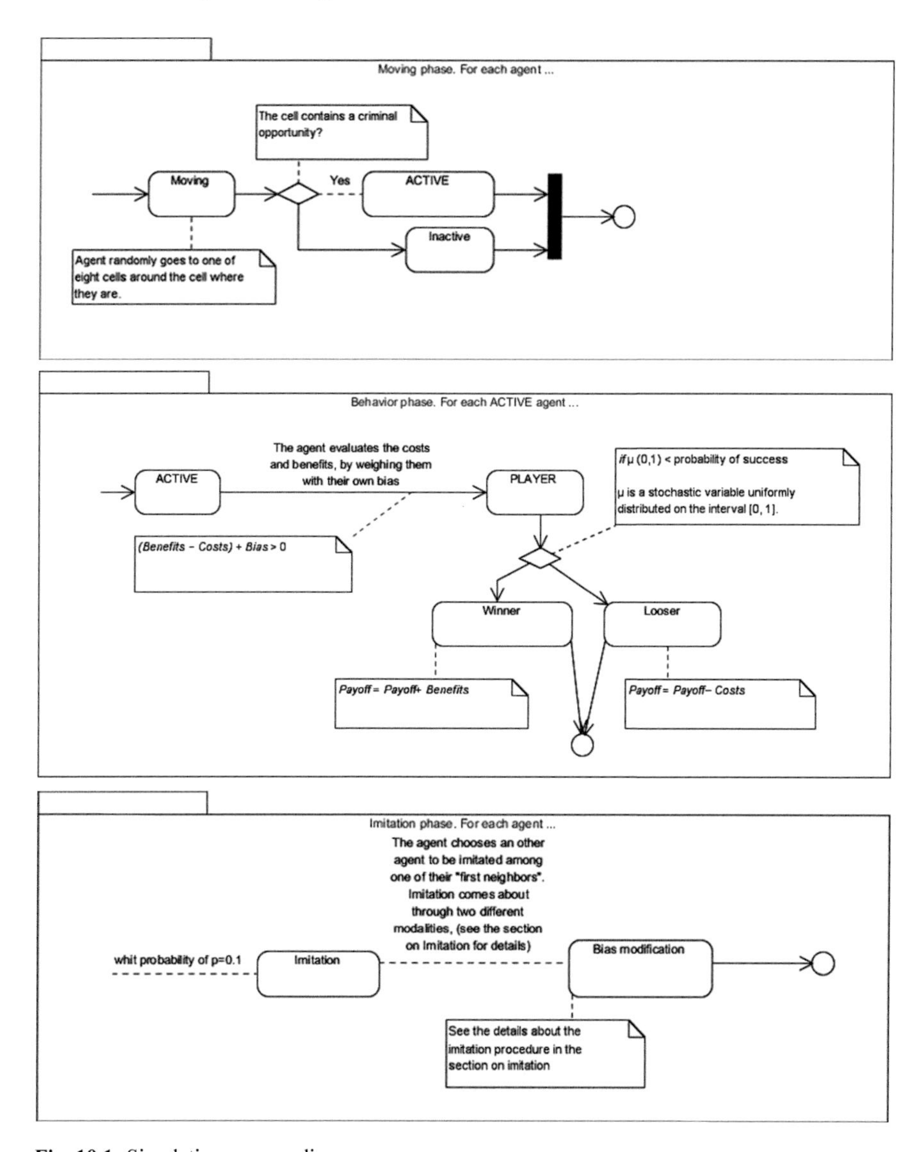

Fig. 10.1 Simulation process diagram

10.4.1 Steps of the Simulation

Each simulation run comprises three steps:

1. All agents move on a bidimensional world divided into cells (regular lattice) and assess if there is an opportunity to commit a crime;
2. Those agents who end up on patches with criminal opportunities decide whether to take them;

3. Agents decide whether to imitate decisional biases of others using the network structure. Each agent imitates one of their first neighbors on the network according to *rational* or *social imitation*;

The simulation process can be summarized as depicted in Fig. 10.1.

As illustrated in the simulation process diagram, after the moving phase in which all agents start moving around the environment encountering some criminal opportunities, the simulation enters a behavior phase in which those agents that encounter an opportunity for crime are allowed to choose whether to take it, making an individual assessment between its costs and benefits, on the basis of their individual bias. As a result, agents decide whether to "play" or not, i.e, whether to take a criminal opportunity or not.

As shown in the behavior phase, the outcome of their choices will have different consequences to them, affecting their level of *payoff*, another individual property, which can increase or decrease. In other words, when an actor undertakes a criminal opportunity they can "win" or "lose": if they win, their payoff increases by the benefit value associated to the criminal opportunity; and on the contrary, if they lose, their payoff decreases by the amount of the costs associated to that criminal opportunity.

We must note that, when summarizing the decision process in the model, there are three factors that determine the agent decision to perform a deviant/criminal action: *Benefits* and *Costs* of criminal opportunities and *individual bias*. First, *Benefit* is an objective property associated with the opportunity to commit crimes and favors the decision to take the opportunity (action = 1); second, *Cost* is an objective property associated with the criminal opportunity and discourages agents from taking the opportunity; third, *Bias* is a subjective property of agents (the individual evaluation of costs and benefits of criminal action made by agents).

The agent's decision process does not take into consideration the *probability of success* associated with criminal opportunities, where agents are not aware of how likely the probability of carrying out the offense successfully is (that is, the probability of winning that is associated with each criminal opportunity).

From the observer/researcher's point of view, knowing the values of the *probabilities of success* associated with criminal opportunities, it is possible to estimate the *Expected payoff* associated with each criminal opportunity (which is different from the real *payoff* reached by agents as the outcome of their choices).

Moreover, knowing the values of *probabilities of success*, it is also possible to estimate those values of *bias* for which agents will make the "convenient"[9] choice according to a specific decision context, that is: they will take criminal opportunities when it is convenient for them to do it (there is a high probability of carrying out the offense successfully) and they will not take criminal opportunities when it

[9] The term *convenient* is used in a classical rational-choice perspective to indicate that a certain choice made by agents suits the decisional conditions. That choice could be convenient from the agent's point of view but not be generally accepted by public opinion or from a moral point of view, as it leads to the violation of law or it is a deviant choice. In other words, convenient is only considered from the rational agent's point of view and not from the researcher's perspective.

is not convenient to do so (there is a high probability of carrying out the offense unsuccessfully).

At each run of the simulation it is then possible to estimate the amount of agents who have made the "convenient" choice (this is measured by an observer parameter called *right behavior* that is useful to the researcher in order to see what happens in the simulation settlement).

At this point the comparison between the two imitation models comes into play. The following section on imitation shows in details the imitation phase of the simulation process.

During the imitation phase agents choose another agent to be imitated, among one of their "first neighbors."[10] Imitation means that the value of the *bias* of the agents (that is their attitude toward crime or, in other words, their perception of criminal opportunities) becomes more similar to those associated with the person being imitated.

For each agent *i* who imitates an agent *j*, imitation comes about according to Eq. (10.1), where B_i is the bias of the agent *i* and B_j is the bias of agent *j*.

$$Bi = (Bj - Bi) * 0.1 + N(0;1) \tag{10.1}$$

A numerical example may clarify how imitation between agents comes about. If an agent *i* with a *bias* value of 0.2 imitates an agent *j* with a *bias* value of 0.5, the bias value of the agent *i* will become more similar to that of the agent *j*, as follows:

$$Bi = (0.5 - 0.2) * 0.1 + N(0;1) \tag{10.2}$$

We assumed that the learning mechanism takes place through two different types of imitation, which we compared. Then, the analytical core of the formal model is the identification of *j*, i.e., the target who is being imitated. We distinguished two methods to identify the target. In fact, different types of imitative behavior differ for the motivations that are behind the imitation, i.e., the person who is being imitated.

The two different types of imitative learning that we tested in our model are:

- *Rational imitation*: this kind of imitative behavior is based on the payoff of the other actors in the network: an actor imitates the behavior of the actor who reached the greatest payoff. More specifically, *rational imitation* is based on the *performance* observed.
- *Social imitation*: this kind of imitative behavior is based on the social prestige of the other actors in the network. More specifically, *social imitation* is based on the degree of connectivity observed: an actor imitates the behavior of the actor who has the highest number of connections in the network.

[10] "First neighbors" means ego's neighbors, that is, all those agents who can be reached by just one step starting from ego. According to the theory of *differential association*, the impact of others (peers, neighbors) on decisions to commit a crime is influenced by the learning processes of imitation.

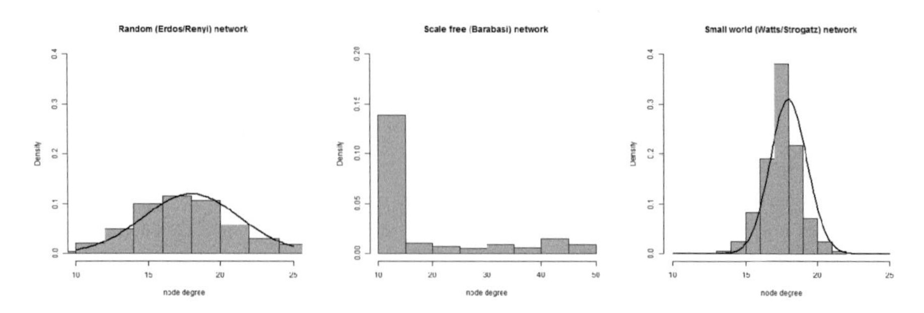

Fig. 10.2 Degree distribution for the three different network topologies

10.4.2 Network Topologies

Agents are linked together in a network and they can only imitate other agents connected to the network. In fact, individuals interact within social networks, influencing each other through the mechanisms of imitation. Therefore we use the network to know who imitate.

We use three different network topologies: random, scale-free, and small world topology[11]. The three different network topologies are characterized by the *degree distribution* shown in Fig. 10.2.

At this point, we may ask whether one of the two mechanisms of imitation allow the *bias* of agents to evolve in such a way that agents will take criminal opportunities only if it is "convenient" for them. We may also question whether *network topologies* affect the evolution of bias *via* imitation.

The purpose of our paper is to explore whether there are any conditions in which these mechanisms of imitation, especially in relation with network topologies, affect individual criminal choices and consequently the spreading of crime pathways.

In the following section we provide the results of our simulation study.

10.5 Simulation Results

We explored the interplay among four independent variables—each of them was observed in different modalities (Table 10.1). Our independent variables were *network topologies*; *average probability of success* (which measures the average probability of winning associated with criminal opportunities displaced in the environment); *quality of the criminal decisional environment* (which measures the difference between *Benefits* and *Costs*); and *mechanisms of imitation.*

[11] The three algorithms used for the creation of the topologies are—the classical Erdos algorithm for Random network (Erdos and Renyi 1959); the preferential attachment algorithm for scale-free network (Barabási and Albert 1999); and the rewiring algorithm for small-world network (Watts and Strogatz 1998).

Table 10.1 Independent variables and modalities

Independent variables	Modalities
Quality	Low (− 1)-High (1)
The average probability of winning associated with the criminal opportunities	Low (0.2)-High (0.8)
Mechanism of imitation	Rational-social
Network topologies	Random-scalefree-smallworlds

If the *average probability of success* is low we can say that the opportunities in the environment are unfavorable and risky, on average. If the *average probability of success* is fair, there is on average the same probability of winning or losing the game.

Quality refers to how the decisional environment appears to the agents, as they do not take into consideration the probabilities of being punished. Therefore, we distinguish decisional environments that are "deceitful" in a positive sense (which means that it is apparently convenient to take criminal opportunities) from those apparently negative.

If *quality* is low (− 1), we can say that, on average, *Costs* of criminal opportunities are higher than *Benefits*. Consequently, the decisional environment appears negative. On the contrary, a high *quality* (1) means that criminal opportunities bring higher benefits then costs. Consequently the decisional environment appears positive.

In our simulation study, we used a logical time unit of *ticks*. A *tick* refers to the time required for an agent to undertake at least one criminal/deviant action. We did a multi-run simulation. For each scenario (each different combination of experimental modalities) we ran 40 different simulations. The total length of each simulation is 10,000 *ticks* (Table 10.1).

We observed the trend of some dependent variables:

1. The *percentage of deviants*: percentage of agents that decide to undertake a criminal/deviant action at each *tick* of simulation;
2. The *right behavior*: the percentage of agents who make the "convenient" choice at each *tick* of simulation.

After a set of experimental runs, we drew some of our simulation results, as follows: The graphs report the average values of the parameters manipulated and the error bar (+/− 1 standard deviation).

The most interesting results emerging from our simulated experiments are the following: Our first experiments compared the *percentage of deviants* resulting from the two imitation modalities by varying the *quality* of the decisional environment and the *probability of success* associated—on average—with criminal opportunities. The simulation results are an average on the three topologies. In fact, in these experimental conditions the cases structure of the topology did not have any effect on the results.

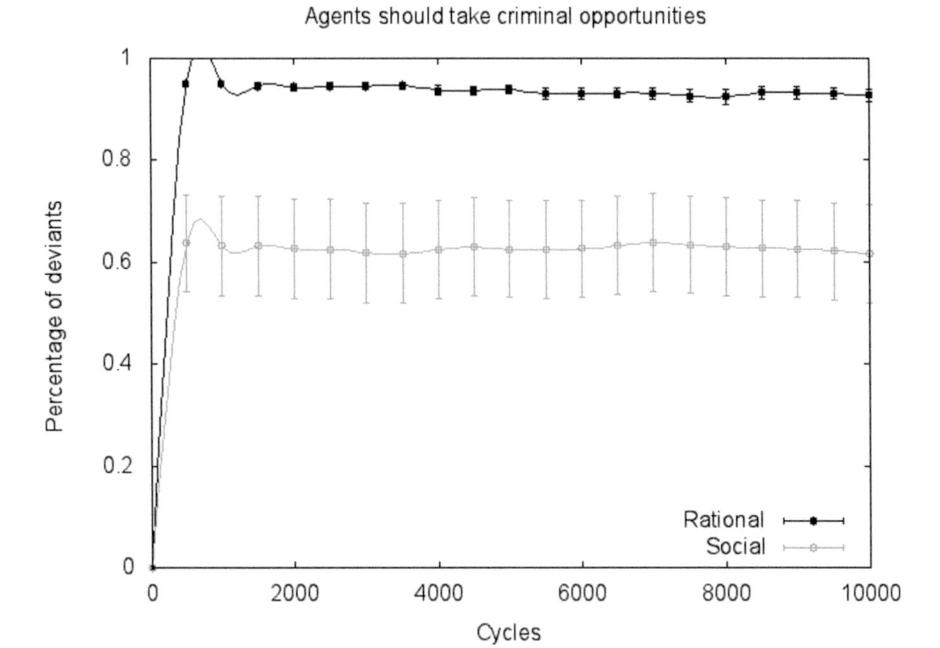

Fig. 10.3 Percentage of deviants generated by rational imitation and social imitation with **a** high quality (1) and a high average probability of success (0.8). **b** low quality (− 1) and a high average probability of success (0.8). In both cases, the expected payoff is greater than zero

Figure 10.3 shows that, when it is convenient for the agents to take criminal opportunities (that is, the *expected payoff* is greater than zero), almost every agent using a rational imitation takes the criminal opportunity, whereas the *percentage of deviants* using social imitation is lower. As a reminder, agent's decision process does not take into consideration the *probability of success* associated with criminal opportunities, where agents decide whether to undertake a deviant/criminal action by performing an evaluation of costs and benefits of the opportunity for crime they encounter. Thus, in this specific decision context, agents using a rational imitation learn quickly to perform the convenient choice (they perform what we called the *right behavior)* whereas agents using social imitation do not learn to make the most convenient choice. In this specific experimental condition the *right behavior* means to take criminal opportunities where, from a researcher's point of view, the *expected payoff* is positive.

As regards the opposite experimental condition, concerning decisional contexts in which it is not convenient for the agents to take the criminal opportunities, as shown in Fig. 10.4, the resulting *percentage of deviants* generated by both rational and social imitation, compared to those shown in Fig. 10.3, is quite different.

By comparing the two graphs (Fig. 10.3 and 10.4), we can see that the *percentage of deviants* using rational imitation varies significantly between the two experimental conditions, despite the fact that they are symmetrical. This difference

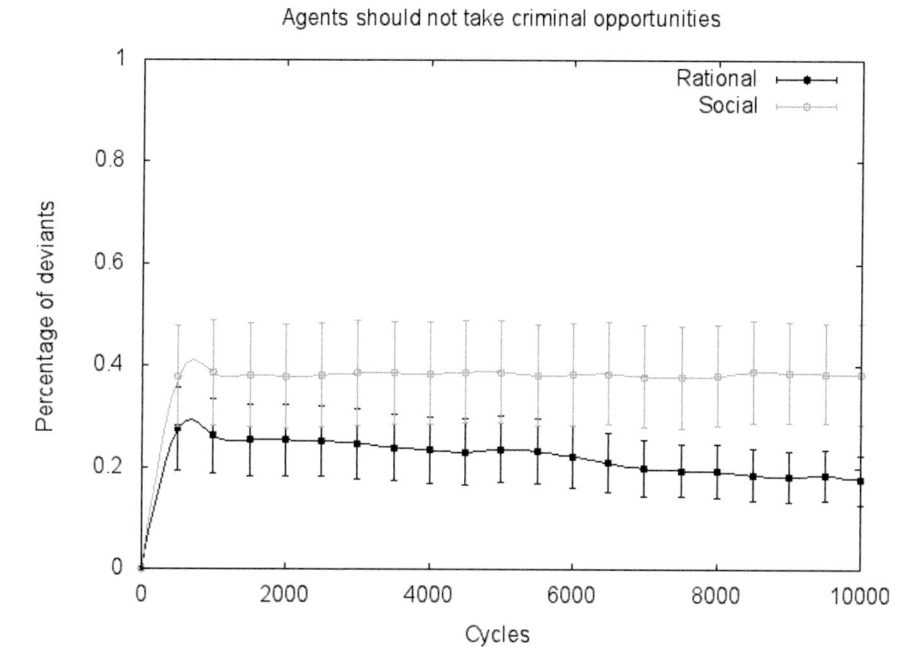

Fig. 10.4 Percentage of deviants generated by rational imitation and social imitation with **a** low quality (− 1) and a low average probability of success (0.2). **b** high quality (1) and a low average probability of success (0.2). In both cases, the expected payoff is lower than zero

relies on the fact that our model, unlike the standard decision-theory model, copies decisions related to criminal behavior. Therefore, we ensured that the agents' *payoff* is not affected by the decision to not take criminal opportunities.

The main evidence resulting from the first set of simulations concerns the effects of rational imitation on individual criminal choices and on the spreading of crime. Rational imitation allows the bias of agents to evolve in such way that they will take the criminal opportunities when it is "convenient" for them, unlike agents using a social imitation modality. Thus, rational imitation affects the spreading of crime across social networks more than social imitation.

Other simulations focused on the comparison between the two opposite "deceitful" experimental conditions (see Fig. 10.5). Note that in our model we distinguished between decisional environments that are "deceitful" in a positive sense from those apparently negative.

By comparing the percentage of agents performing the *right behavior*—according to a rational imitation mechanism—in the two opposite "deceitful" contexts, some differences emerge, despite the fact that—once again—they are symmetrical.

Specifically, when the decisional environment appears positive to the agents, but actually it is not convenient for them to take criminal opportunities, agents using rational imitation perform the *right behavior* less often than when the "deceitful"

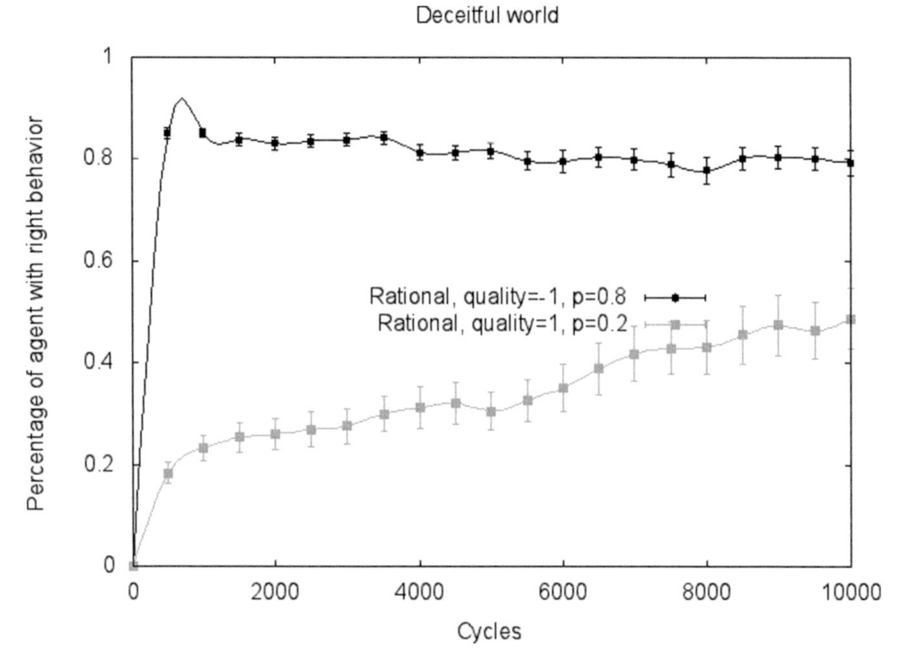

Fig. 10.5 Percentage of agents with right behavior using rational imitation in both positive and negative "deceitful" decisional contexts: **a** low quality (−1) and a high average probability of success (0.8). **b** high quality (1) and a low average probability of success (0.2)

context appears negative to them. In the latter case, agents do learn *via* rational imitation to perform the so called *right behavior*.

Thus, the two "deceitful" contexts affect the spread of crime through rational imitation differently, despite the fact that they are symmetrical. From a sociological point of view, this evidence suggests that the behavior of the agents evolves *via* bias imitation when the decisional environment is beneficial for them, despite their perceptions. On the contrary, when it is apparently convenient to take criminal opportunities, it will take more time to learn to refrain from committing crimes.

Further experiments focused on comparing the three network topologies (see Fig. 10.6). We did 40 simulation runs for each network topology in order to observe the dynamics of the *percentage of deviants*. Specifically, we observed the *percentage of deviants* generated through social imitation, by maintaining a constant high *probability of success* and high *quality*. In this experimental condition, the decisional environment is not "deceitful," whereas it appears positive to the agents, and actually it is convenient to take criminal opportunities.

Our simulations showed that the *percentage of deviants* resulting from the three network topologies was very different.

In this experimental condition agents' decisional process normally leads them to take the criminal opportunities they encounter, where it is convenient to do so (the *expected payoff* is positive). As we can see in Fig. 10.6, when the learning mecha-

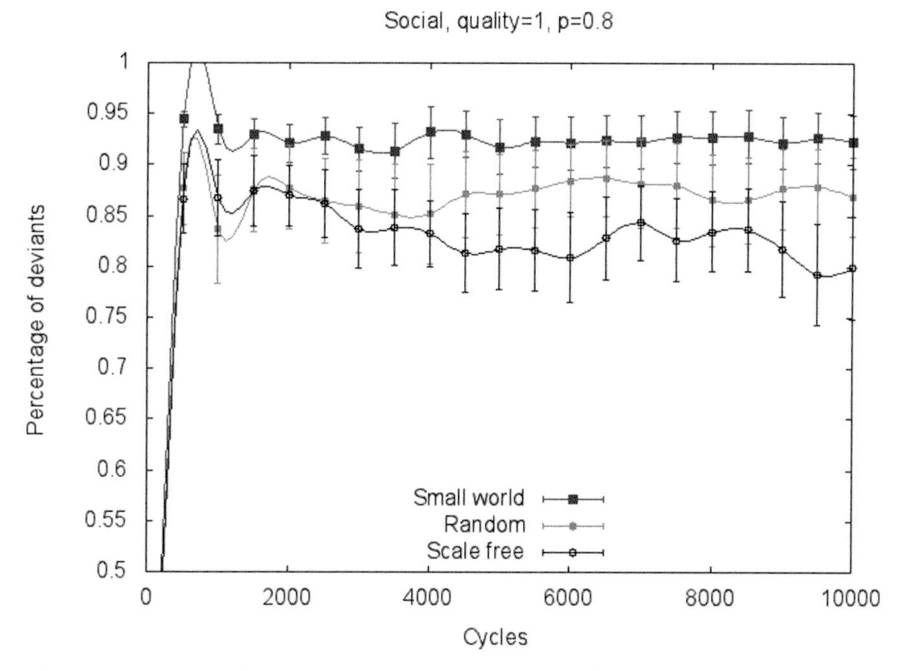

Fig. 10.6 Percentage of deviants in the three network topologies with: social imitation—high quality (1) and high average probability of success (0.8)

nism in social networks takes place through a social imitation, the network topology affects the spreading of crime.

Specifically, as far as the small-world network is concerned, as can be seen in Fig. 10.6, the *percentage of deviants* curve is higher than those of both random and scale-free networks. This means that, most frequently, agents interacting in a small-world network through social imitation make their decision whether to undertake a deviant/criminal action on the basis of a general *risk propensity*, which pushes people toward antisocial or criminal behavior.

On the contrary concerning the scale-free network the *percentage of deviants* curve is lower. It means that scale-free networks allow the spreading of a general *risk aversion*, which prevents agents from taking criminal opportunities.

10.6 Discussion

As ascertained by Sutherland onward, learning processes through imitation on social networks are the basis of social influence. Despite their importance, the causal mechanisms that link imitation to crime are still poorly understood. In fact, neither of the current approaches on imitation seems to provide an explicit *explanatory mechanism* underlying imitative behavior (Laland and Bateson 2001).

Recently, criminological studies have explored the impact of social influence and imitation on crime, devoting a growing focus to conceptualizing criminal groups as networks (Ormerod and Wiltshire 2009). Building on these scenarios, our study aimed to explore the *casual mechanisms* underlying imitation and their effects on crime. In other words, we were interested in investigating whether and how imitation, especially in relation with *social network topologies*, affects the spread of crime pathways.

We hypothesized different criminal outcomes generated by different learning mechanisms of imitation at the micro-level of social interaction. Specifically, we distinguished between *rational imitation* (Hedström 1998; Schwier et al. 2004) and *social imitation*, on the basis of the motivations that are behind the imitative behavior. *Rational imitation* is based on the *performance* observed, where the likelihood that a model's behavior will be imitated is contingent on its observed *consequences* (Akers and Jennings 2009, p. 109). Conversely, *social imitation* is based on *social prestige* acquired by those agents who are strongly embedded into the social network and who perform the function of "socialization to deviance" (Becker 1963, 1967).

In order to test our hypothesis, we developed an agent-based theoretical model which allowed us to formalize the structure of different types of potential networks— random, scale-free and small-world—on which agents are connected, as well as to investigate the effects of the two mechanisms of imitation.

Results from the simulations reveal the impact of the mechanisms of imitation in producing the spreading of crime and the role of *network topologies*. The main substantive implications emerging from our simulation study concern the different effects of rational and social imitation on crime. Agents' behavior, through a rational imitation, seems to evolve in such way that agents undertake a criminal choice when it is beneficial for them, unlike agents using a social imitation. Thus, from a sociological perspective we may argue that criminal patterns that have been rewarded flow across social networks through a learning mechanism based on rational imitation.

Some other sociological issues concern *how* rational imitation comes about on social networks. Specifically, we found out that the rational imitation mechanism is affected by how the decisional context appears to the agents. In fact, when the decisional environment appears positive to the agents, they are prone to take criminal opportunities. In this case, rational imitation does not allow the behavior of agents to *evolve*; that is, agents do not learn to do the convenient choice. In other words, in those conditions, the agents do not learn, via rational imitation, to refrain from committing crimes. This evidence is in accordance with the theoretical statement, widespread in criminology, that individuals are usually more attracted by the benefits of crime and are less willing to take into considerations the costs of crime (see Gottfredson and Hirschi 1990). Consequently, if agents perceive the context as favorable to crime it will take more time for them to learn to refrain from being involved in it.

This evidence relies on the structure of the rational imitation mechanism, based on a costs–benefit evaluation of the observed consequences of behavioral models, and then affected by the so called *heuristics* and *biases* that, from a sociological perspective, characterize individual decision-making (Elster 1999; Boudon 2003). Specifically, individual attitude toward risks, and the consequent decision to be engaged in crime, are influenced by the way in which the "prospects of choice" appear to the agents, where benefits and costs have a different subjective utility (see also Kaheneman and Tversky 1979). Thus, the way in which risky alternatives are framed affects how the rational imitation mechanism takes place.

Finally, our simulation model sheds light on the effects of the three network topologies on the spreading of crime. Specifically, we are able to deduce from our agent-based model the type of social network across which crime flows and how agents influence each other's behavior.

First of all, the network topology seems to have some effect on crime when individuals influence each other through social imitation. In such cases, agents imitate other actors connected on their network, on the basis of their degree of connectivity, which marks *social prestige*. We have pointed out the *social-relationship-oriented* or so called *conformity motivations* which acquire a sociological relevance in the explanation of imitative behavior. From this perspective, social imitation relies on the motivation to affiliate with others as well as on *social power* (Cialdini 2004, p. 595).

Secondly, our results suggest that the relevant network to account for the spreading of crime through social imitation has a small-world structure. We show that crime appears to flow across a small-world network. In fact, our model suggests that agents interacting in small-world networks, through social imitation, maintain a propensity toward risks which favors behaviors towards crime. This result confirms the hypothesis of a "contagion" effect triggered by the specific structure of the small-world network, which resembles the structure of overlapping groups of "friends of friends" (Ormerod and Wiltshire 2009).

Thus, in this kind of social context, social imitation mechanism based on conformity motivations among friendships networks generates a "contagion" effect which accounts for the observed spreading of crime pathways.

References

Akers, R. L. (1985). *Deviant behavior: A social learning approach* (3rd ed). Belmont: Wadsworth.

Akers, R. L. (1998). *Social learning and social structure: A general theory of crime and deviance.* Boston: Northeastern University Press.

Akers, R. L., & Jennings, W. G. (2009). The social learning theory of crime and deviance. In M. Krohn, A. Lizotte, & G. Hall. (Eds.), *Handbook on crime and deviance* (pp. 103–120). Springer.

Akers, R. L., & Jensen, G. F. (2006). The empirical status of social learning theory of crime and deviance: The past, present, and future. In F. T. Cullen, J. P. Wright, & K. R. Blevins (Eds.), *Taking stock: The status of criminological theory.* New Brunswick: Transaction Publishers.

Akers, R. L., & Sellers, C. S. (2009). *Criminological theories: Introduction, evaluation, and application* (5th ed.). Losangeles: Roxbury Publishing.

Bandura, A. (1977). *Social learning theory.* New York: General Learning Press.

Barabási, A. L., & Albert, R. (1999). Emergence of scaling in random networks. *Science, 286*(5439), 509–512.
Becker, H. S. (1963). *Outsiders: Studies in the sociology of deviance*. New York: Free Press.
Becker, H. S. (1967). History, culture and subjective experience: An exploration of the social bases of drug-induced experiences. *Journal of Health and Social Behavior, 8*(3), 163–177.
Birks, D. J., Donkin, S., & Wellsmith, M. (2008). Synthesis over analysis: Towards an ontology for volume crime simulation. In L. Liu & J. E. Eck (Eds.), *Artificial crime analysis systems* (pp. 160–192). Hershey: IGI Global.
Bosse, T., Gerritsen, C., & Klein, M. C. A. (2009). Agent-based simulation of social learning in criminology. Proceedings of the "International Conference on Agents and Artificial Intelligence" (ICAART), Porto, Portugal, 2009.
Boudon, R. (2003). Beyond rational choice theory. *Annual Review of Sociology, 29*, 1–21.
Brezina, T., & Piquero, A. R. (2003). Exploring the relationship between social and non-social reinforcement in the context of social learning theory. In R. L. Akers & G. F. Jensen (Eds.), *Social learning theory and the explanation of crime: A guide for the new century. Advances in criminological theory* (Vol. 11, pp. 265–288). New Brunswick: Transaction Publishers.
Bruinsma, G., & Bernasco, W. (2004). Criminal groups and transnational illegal markets. A more detailed examination on the basis of Social Network Theory. *Crime, Law & Social Change, 41*, 79–94.
Burgess, R., & Akers, R. L. (1966). A differential association-reinforcement theory of criminal behavior. *Social Problems, 14*, 363–383.
Calvo-Armengol, A., & Zenou, Y. (2004). Social networks and crime decisions: The role of social structure in facilitating delinquent behavior. *International Economic Review, 45*, 939–958.
Chappell, A. T., & Piquero, A. R. (2004). Applying social learning theory to police misconduct. *Deviant Behavior, 25*, 89–108.
Chartrand, T. L., & Bargh, J. A. (1999). The chameleon effect: The perception-behavior link and social interaction. *Journal of Personality and Social Psychology, 76*, 893–910.
Cialdini, R. B. (2001). *Influence: Science and practice* (4th ed.). Boston: Allyn & Bacon.
Cialdini, R. B., & Goldstein, N. J. (2004). Social influence: Compliance and conformity. *Annual Review of Psychology, 55*, 591–621.
Clarke, R. V. (1997). *Situational crime prevention: Successful case studies* (2nd ed.). Albany: Harrow & Heston.
Cohen, L. E., & Felson, M. (1979). Social change and crime rate trends: A routine approach. *American Sociological Review, 44*, 588–607.
Cornish, D. B., & Clarke, R. V. (1987). Understanding crime displacement: An application of rational choice theory. *Criminology, 25*, 933–947.
Elliott, D. S., Wilson, W. J., Huizinga, D., Sampson, R. J., Elliott, A., & Rankin, B. (1996). The effects of neighborhood disadvantage on adolescent development. *Journal of Research in Crime and Delinquency, 33*, 389–426.
Elster, J. (1999). *Alchemies of the mind: Rationality and the emotions*. Cambridge: Cambridge University Press.
Erdos, P., & Renyi, A. (1959). On random graphs. *Publicationes Mathematicae, 6*, 290–297
Felson, M. (2002). *Crime and everyday life* (3rd ed.). Thousand Oaks: SAGE publications.
Furtado, V., Melo, A., Coelho, A., Menezes, R., & Belchior, M. (2008). Simulating crime against properties using swarm intelligence and social networks. In L. Liu & J. E. Eck (Eds.), *Artificial crime analysis systems* (pp. 300–319). Hershey: IGI Global.
Gilbert, N. (2008). *Agent-based models*. London: Sage.
Gilbert, N., & Troizsch, K. (2005). *Simulation for the Social Scientists*. Buckingham-Philadelphia: Open University Press.
Glaeser, E., Sacerdote, B., & Scheinkman, J. A. (1996). Crime and social interaction. *The Quarterly Journal of Economics, CXI*, 507–548.
Gottfredson, M., & Hirschi, T. (1990). *A general theory of crime*. Stanford: Stanford University Press.

Groff, E. (2007). Simulation for theory testing and experimentation: An example using routine activity theory and street robbery. *Journal of Quantitative Criminology, 23*(2), 75–103.

Groff, E. (2008a). Characterizing the spatio-temporal aspects of routine activities and the geographic distribution of street robbery. In L. Liu & J. E. Eck (Eds.), *(2008). Artificial crime analysis systems: Using computer simulations and geographic information systems* (pp. 226–251). Hershey: IGI Global.

Haynie, D. L. (2002). Friendship networks and delinquency: The relative nature of peer delinquency. *Journal of Quantitative Criminology, 18*, 99–134.

Hedström, P. (1998). Rational imitation. In P. Hedström & R. Swedberg (Eds.), *Social mechanism* (pp. 306–328). Cambridge: Cambridge University Press.

Hedström, P. (2005). *Dissecting the social: On the principles of analytical sociology*. Cambridge: Cambridge University Press.

Hedström, P., & Åberg, Y. (2006). Ricerca quantitativa, modelli ad agenti e teorie del sociale. In P. Hedström (Ed.), *Anatomia del Sociale* (pp. 143–179). Milano: Bruno Mondatori.

Hedström, P., & Swedberg, R. (1998). *Social mechanisms. An analytical approach to social theory*. Cambridge: Cambridge University Press.

Heyes, C. M., & Ray, E. D. (2000). What is the significance of imitation in animals? *Advances in the Study of Behavior, 29*, 215–245.

Hurley, S., & Charter, N. (2005). *Perspectives on imitation: From neuroscience to social science* (Vol. 2). Cambridge: MIT Press (in multiple jurisdictions).

Kaheneman, D., & Tversky, A. (1979). Prospect theory: An analysis of decision under risk. *Econometrica, 47*, 263–291.

Katz, L., Kling, A., & Liebman, J. (2001). Moving to opportunity in Boston: Early results of a randomized mobility experiment. *Quarterly Journal of Economics, CXVI*, 607–654.

Kaza, S., Xu, J., Marshall, B., & Chen, H. (2005). Topological analysis of criminal activity networks. Proceedings of the 2005 national conference on Digital government research, pp. 251–252.

Lakin, J. L., & Chartrand, T. L. (2003). Using nonconscious behavioral mimicry to create affiliation and rapport. *Psychological Science: A Journal of the American Psychological Society/APS, 14*, 334–339.

Laland, K., & Bateson, P. (2001). The mechanisms of imitation. *Cybernetics and Systems: An International Journal, 32*, 195–224.

Liu, L., & Eck, J. E. (Eds.). (2008). *Artificial crime analysis systems: using computer simulations and geographic information systems*. Hershey: IGI Global.

Liu, L., Wang, X., Eck, J., & Liang, J. (2005). Simulation crime events and crime patterns in RA/CA model. In F. Wang (Ed.), *Geographic information systems and crime analysis* (pp. 197–213). Singapore: idea Group.

Lowe, M. (2009). Attachment mechanisms in catalogue networks, working paper.

Ludwig, J., Hirschfeld, P., & Duncan, G. (2001). Urban poverty and juvenile crime: Evidence from a randomized housing-mobility experiment. *Quarterly Journal of Economics, CXVI*(2), 665–679.

Manski, C. (1993). Identification problems in social sciences. *Sociological Methodology, 23*, 1–56.

Manski, C. (2000). Economic analysis of social interaction. *Journal of Economic Perspectives, 14*(3), 115–136.

Manzo, G. (2004). Appunti sulla simulazione al computer. Un metodo per la ricerca sociologica. In C. Corposanto (Ed.), *Metodologie e tecniche non intrusive nella ricerca sociale*. Milano: FrancoAngeli.

Manzo, G. (2007). Variables, mechanisms, and simulations: Can the three methods be synthesized? A critical analysis of the literature. *Revue Francaise de Sociologie-An Annual English Selection, 48*(Suppl.), 35–71.

Manzo, G. (2013). Educational choices and social interactions: A formal model and a computational test. *Comparative Social Research, 30*, 47–100.

McCarthy, B. (2002). New economics of sociological criminology. *Annual Review of Sociology, 28*, 417–442.

Mehlkop, G., & Graeff, P. (2010). Modelling a rational choice theory of criminal action: Subjective expected utilities, norms and interactions. *Rationality and Society, 22*(2), 189–222.

Meltzoff, A. N. (1995). Understanding the intentions of others: Re-enactment of intended acts by 18-month-old children. *Developmental Psychology, 31*, 838–850.

Miller, J. H., & Page, S. (2007). *Complex adaptive systems. An introduction to computational models of social life*. Princeton-Oxford: Princeton University Press.

Nagin, D. S. (1998). Criminal deterrence research at the outset of the twenty-first century. In M. Tonry (Ed.), *Crime and justice, a review of research* (Vol. 23, pp. 1–42). Chicago: University of Chicago Press.

Ormerod, P., & Wiltshire, G. (2009). Binge' drinking in the UK: A social network phenomenon. *Mind & Society, 8*, 135–152.

Sampson, R. J., Raudenbush, S. W., & Earls, F. (1997). Neighborhoods and violent crime: A multilevel study of collective efficacy. *Science, 277*, 918–924.

Sawyer, R. K. (2003). Artificial societies: Multi-agent systems and micro-macro link in sociological theory. *Sociological Methods & Research, 31*(3), 325–363.

Scheinkman, J. A. (2008). Social interaction, working paper.

Sellers, C. S., Cochran, J. K., & Winfree, T. L. (2003). Social learning theory and courtship violence: An empirical test. In R. L. Akers & G. F. Jensen (Eds.), *Social learning theory and the explanation of crime: A guide for the new century. Advances in criminological theory* (Vol. 11, pp. 109–128). New Brunswick: Transaction Publishers.

Sutherland, E. H. (1947). *Principles of criminology* (4th ed.). Philadelphia: J. B. Lippincott.

Sutherland, E. H., & Cressey, D. R. (1966). *Principles of criminology* (7th ed.). Philadelphia: J.B. Lippincott.

Triplett, R., & Payne, B. (2004). Problem solving as reinforcement in adolescent drug use: Implications for theory and policy. *Journal of Criminal Justice, 32*, 617–630.

Wang, X. (2005). Spatial adaptive crime event simulation With RA/CA/ABM computational laboratory, Geography, Cincinnati: University of Cincinnati.

Wang, X., Liu, L., & Eck, J. E. (2008). Crime simulation using GIS and artificial intelligent agents. In L. Liu & J. E. Eck (Eds.), *Artificial crime analysis systems* (pp. 209–225). Hershey: IGI Global.

Warr, M. (2002). *Companions in crime: The social aspects of criminal conduct*. Cambridge: Cambridge University Press.

Watts, D. J., Strogatz, S. H. (1998). Collective dynamics of 'small world' networks. *Nature, 393*, 440–442

Wikström, P.-O. H. (2004). Crime as alternative: Towards a cross-level situational action theory of crime causation. In J. McCord (Ed.), *Beyond empiricism: Institutions and intentions in the study of crime. Advances in criminological theory* (Vol. 13, pp. 1–37). New Brunswick: Transaction.

Wikström, P.-O. H. (2006). Individuals, settings and acts of crime: Situational mechanisms and the explanation of crime. In P.-O. H. Wikström & R. J. Sampson (Eds.), *The explanation of crime: Context, mechanisms and development*. Cambridge: Cambridge University Press.

Wikström, P.-O. H. (2010). Situational Action Theory. In F. Cullen & P. Wilcox (Eds.), *Encyclopedia of criminological theory*. London: SAGE Publications.

Wikström, P.-O. H., & Treiber, K. (2007). The role of self-control in crime causation: Beyond Gottfredson and Hirschi's general theory of crime. *European Journal of Criminology, 4*(2), 237–264.

Wikström, P.-O. H., & Treiber, K. (2009). Violence as situational action. *International Journal of Conflict and Violence, 3*(1), 75–96.

Wilensky, U. (1999). Netlogo, http://ccl.northwestern.edu/netlogo/, Center for Connected Learning and Computer-Based Modeling, Northwestern University, Evanston, IL.

Chapter 11
NewsMarket 2.0: Analysis of News for Stock Price Forecasting

Alessandro Barazzetti and Rosangela Mastronardi

11.1 Introduction, Motivation, and Related Literature

Web news can be used to accurately track not only several social phenomena but also financial trends (Choi and Varian 2009; Preis et al. 2013). Financial market systems are complex, and therefore trading decisions are usually based on information about a huge variety of socioeconomic topics and societal events.

Predicting stock trends has long been an intriguing topic and has been extensively studied by researchers from different fields. There is a large literature about how macroeconomic news influences market; and investors, economists, and journalists follow monthly macroeconomic data releases concerning economic conditions. A huge problem with this data, however, is that the information is available with a lag that increases significantly very quickly. In fact, the data for a given month are generally released about halfway through the next month and are typically revised a few months later: this is the reason that over the last few years a new approach, based on machine learning, has been extensively studied for its potential to predict the direction of financial markets and give answers to questions like, "What is the consequence of the civil war in Libya for financial markets?" (Casti 2012; Scheffer 2009).

Applying the theory and tools of information and communications technology (ICT), statistics, and econometrics, systems were developed that are able to gather and analyze large amounts of data (big data) that is available free on the net (open data). For example, sentiment analysis of the opinion expressed by users of social networks is a closely studied field of application of natural language processing (NLP) to financial markets: this kind of research utilizes tools, useful in automated trading, where machine learning also has a key role.

A. Barazzetti (✉) · R. Mastronardi
QBT, Sagl, Via E. Bossi 4, 6830 Chiasso, Switzerland
e-mail: alessandro.barazzetti@qbt.ch

R. Mastronardi
e-mail: rosangela.mastronardi@qbt.ch

F. Cecconi (ed.), *New Frontiers in the Study of Social Phenomena*,
DOI 10.1007/978-3-319-23938-5_11

The common point of view in all these methods of analysis is the statistical elaboration of texts to extract knowledge about correlation of data: this was not the path we followed. We have developed a model for forecasting stock market trends in different time frames based on the analysis of news events related to a specific domain, in our case the oil and gas markets (Barazzetti et al. 2014), in which the human component of text analysis is predominant in defining the key concepts of the ontology. We investigated whether we could identify key events, through online news, that could be linked to the sign of subsequent stock market moves.

Recently a host of firms, including start-ups as well as established media giants, have been offering tools and services that mine internet data and provide Wall Street with sentiment analysis (Mittermayer 2004; Schumaker and Chen 2006). While there have been several studies covering textual financial predictions, our idea was to collect news or browse e-newspaper sites and before semantically naming the extracted data, to define a priori some correlation with human analysis. Local or global events such as the Ukraine crisis, natural disasters, ruble devaluation, or Brent downward trends, can generate local or global effect in a specific financial sector.

This motivated us to analyze in more detail online news in order to find hidden correlations between news and events in the financial sectors. Tracking down events that signal or anticipate crises, financial turnovers, and financial contagion is of great interest to analysts, investors, and policymakers (Bouchaud 2009) because it may allow for a more prompt portfolio intervention.

Although we restrict our work to stock market moves, our methodology can be readily extended to other domains with macro-economic tendencies. Hence, in contrast to the several prototypes that predict short-term market reactions to news, NewsMarket[1] attempts to forecast medium to long trends of the major equity indices. More precisely, NewsMarket 2.0 is a system, based on human and machine coding, to support decision-making in financial markets (Grimmer and Stewart 2013; Jonathan et al. 2009).

Despite the existence of multiple systems in this area of research that are characterized by the same components—dataset crawling, machine pre-processing and machine learning—our approach in NewsMarket also includes a human element in the second phase of the process. There are at least two reasons that human input is beneficial: the human mind is able to discover hidden correlations better then a machine (even if it is slower than a machine) and to extract a hypothesis from an incomplete set of data using intuition as well as with logical deduction (even if human input is time-consuming and error-prone).

In contrast to other predication tools, once the input data (textual data) is available, it can be fed into a machine-learning algorithm that transforms unstructured text into a representative format that is structured and can be processed by the machine. Human work consists of feature selection (in terms of correlation and length of impact), dimensionality reduction (find and tag key events), and feature representation. In our project human work is very important, because the decision about

[1] NewsMarket is a prototype created by QBT Sagl.

which features will represent a piece of text is crucial: if the representational input is incorrect, nothing more than a meaningless output can be expected.

Having a limited number of features is extremely important, as the increase in the number of features (which can easily happen in feature selection) can make the classification or clustering problems extremely hard to solve, by decreasing the efficiency of most of the learning algorithms. This situation is widely known as the curse of dimensionality (Pestov 2013). In our case, in 2011we identified the first key events with the highest weights as the features, instead of including available concepts, and examined the correlations among them and the stock or sector. We continue to increase our map every year, finding on mean five key events per year. This means that we have created an in-house dictionary or thesaurus dynamically based on the text corpus using term extraction by humans (and not with a inanimate tool) that we call NewsVoc (Soni et al. 2007).

The NewsMarket system does not add anything new to the fields themselves. Its contribution is the creation of the News Index Map (NIM) system, which weaves together disparate fields in the pursuit of solving a discrete prediction problem.

In this chapter we provide a detailed analysis on a particular application of this tool: that is, the anticipation of Eni (an international energy company) trends. In particular, based on the NIM developed in 2011, we show the 2014 and 2015 results.

Our method is capable of automatically identifying events that characterized trends before stock market moves. Below we investigate in detail how today's news about financial stocks and non-financial news, suitably analyzed, can be used as financial indicators for the next month and quarter.

11.2 NewsMarket

Figure 11.1 shows the main architecture of the NewsMarket system and the connections among its three principal components: News Index Map (NIM), the semantic set of concepts (NewsVoc), and the Financial Prompter.

The system's workflow expects that the user queries the system, asking for an investment recommendation about Eni or the oil and gas sector on a daily basis throughout 2015. Once the query has been received, NewsMarket loads all the rules that must be checked in order to provide a recommendation within the framework model of the NewsVoc. The NewsVoc model is based on a previous analysis that assigned to each news/event the NIM for the necessary data.

NewsMarket performs a continuous crawl on the web to gather news about a company or sector for. Once the information has arrived, it will perform a Natural Language Process (NLP) to make interchange calls with the NewsVoc that is in charge of assigning the NIM values. The NewsVoc will finally write the financial reasoning ontology with all the information generated by each rule: this ontology will be sent to the selected component (Financial Prompter). The result of that inference will be processed and returned to the user in an investment answer format. The main components and features of NewsMarket are depicted in Fig. 11.1.

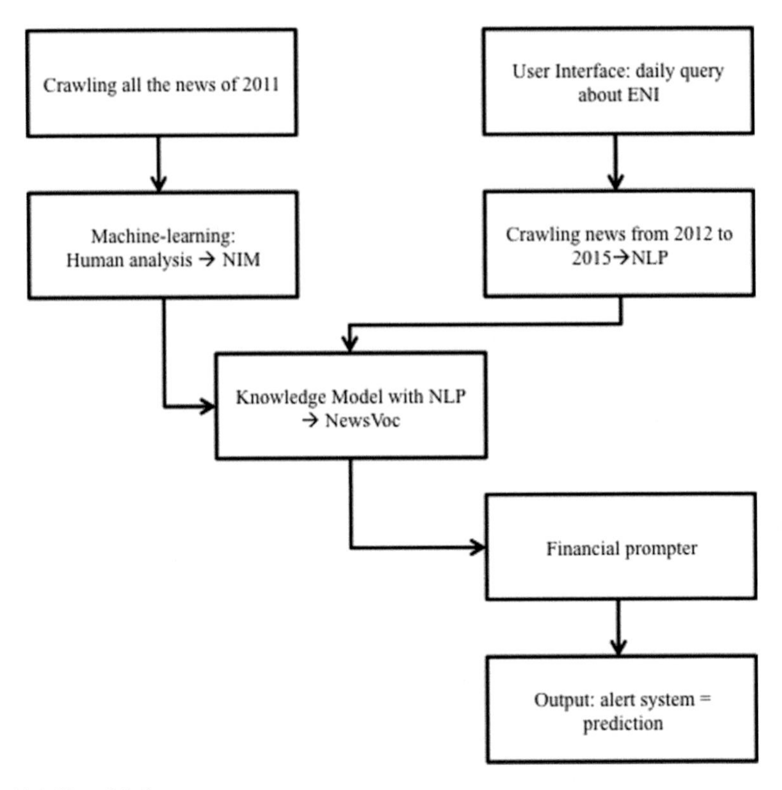

Fig. 11.1 NewsMarket structure

11.2.1 News Index Map (NIM)

The News Index Map (NIM) is the first component of NewsMarket. Here, human experts analyze online news, observe its effect on stocks, and assign, if possible, each event to a specific category (for example war, tsunami, change of CEO, insider trading, etc.).

All the news is read and analyzed by humans, even though this process is time- and resource-intensive. The human element is important to create an appropriate events database[2]. More precisely, humans can classify keywords with the right connotation and assign the correct polarity (bad news or good news). Sometimes, many words may have a negative connotation in one context and a positive connotation

[2] Xie et al. (2008) write: Events can be defined as real-world occurrences that unfold over space and time. In other words, an event has a duration, occurs in a specific place, and typically will involve certain change of state. Using this definition, "a walk on the beach", "the hurricane of 2005", and "a trip to Santa Barbara" would all qualify as events. Events are useful because they help us make sense of the world around us by helping to recollect real-world experiences (e.g., university commencement 2006), by explaining phenomena that we observe (e.g., the annual journey of migrating birds), or by assisting us in predicting future events (e.g., the outcome of a tennis match).

in another. For example, the term *crude oil* may have a negative connotation in an article concerning an oil tanker that crashed and a positive connotation in an earnings report.

Human coders are used also to assigns weights in terms of events' impacts the share prices to evaluate the timing of these effects and the correlation of the news with the referral security or stock sectors.

The extent of prediction between financial and non-financial news articles and their impact on stock market prices is a complex avenue to investigate. While the information contained in financial news articles can have a visible impact on a security's price, information contained in general news that can cause a sudden price movement is more difficult to capture and analyze.

The first challenge of financial and non-financial prediction is to process a large amount of text (in our case for the energy sector) and to select the most suitable websites to crawl for news.

For our study, we used in this phase only Italian sources. One of the principal sources used for the analysis is *Sole24Ore.com*, which offers free real-time and subscription-based services. For the news related to Eni and the energy sector we use *Borsainside.com*.

For each web site source the lastest news is obtained and stored in a database. The information that is retrieved from each news article is the date of publication, the information source, the Url, and the abstract. Abstracts constitute the corpus from which the system extracts the information.

Once all the news that is not linked with Eni is removed, human agents proceed to read the news and assigns weights (in terms of impacts on share prices) and evaluate the timing of the events' effects and the correlation of the news with the stock sector of interest, based on the correlation defined initially by the human agent (see Table 11.1). Hence, we identify:

- The type of correlation (-1 means not the event is not correlated to the securities; 1, otherwise)
- The importance of the event, which is function of the time: (High (H), the news has an immediate effect on securities (1 day); Medium (M), the new has an impact on stocks during the medium term (2–30 days); Low (L), the event will probably have an effect over a month)
- The main objective is to classify the set of news obtained in the previous module according to its polarity: positive (G, Good), negative (N, Negative) or neutral (IN, In line).
- Keywords—selected from the article in well-defined macro-categories (for example Eni, War, Macroeconomic news, Ethical conflicts, etc. …)

In this large amount of data we identify vectors (see Sect. 2.2) of words that generate the same behavioral schema.

We can generalize the behavior of the trend in a matrix of weights 3×3; each single news event can be seen as a sequence of words that has a consequence for the referral stock. This consequence is the combination of the elements of the matrix, which we call NIM: it is the measure of the news and it is a scalar.

Table 11.1 Classification of news

Id_NewsVoc	Weight	NewsVoc	HML	BGIL	C	Word1	Word2	Word3	Word4	Word5	Word6	Word7	Word8	Word9
1	5	tag_acq_coll_accord-tag_brasile-tag_eni-tag_operatore	H	G	1	trattativa	cedere	galp	Petrobas	eni				
12	3	tag_operatore-tag_rating-tag_usa	H	IL	1	aumentare	rating	exxon						
13	3	tag_quotazione_petrolio	H	B	1	diminuire	petrolio	prezzo						
14	3	tag_afrca-tag_page	H	G	1	africa	elezione	nigeria						
23	3	tag_crisi_economica-tag_italia	H	B	-1	consumo	diminuire	Famiglia						
24	8	tag_eia-tag_quotazione_petrolio	H	G	1	aumentare	prezzo	petroio	eia	aumentare	Produzione	Petrolio	eia	
25	4	tag_crisi_debito	H	G	-1	anno	svolta	città	detroit					
26	3	tag_industria_auto-tag_italia	H	G	-1	aumentare	fiat	crysler						
27	4	tag_ribellione_nordafrica	M	B	-1	attacco	infrastruttura	oleodotto	tunisia					
28	4	tag_crisi_debito	M	G	-1	stretta	monetaria	inefficace	cina					
29	10	tag_eia-tag_quotazione_petrolio	H	G	1	diminuire	scorta	petrolio	eia	usa	prezzo	petrolio	aumentare	eia
30	11	tag_quotazione_petrolio	H	B	1	prezzo	petrolio	rialzo	diminuire	scorta	petrolio	aumentare	paura	scompenso
31	3	tag_crisi_debito	M	G	-1	bce	draghi	scegliere						
32	1	tag_crisi_debito	M	G	-1	bond								
33	3	tag_ribellione_nordafrica	M	G	-1	libia	risoluzione	onu						

Hence, NIM as a function of the following three elements:

$$Nim = f\{g(Correlation), k(Timing), j(Effect)\}$$

Where g, k, j are nonlinear functions. Each function may be expressed by one of the following value:

- Correlation (g) can be: + 1 (positive correlation) or − 1 (negative correlation);
- Timing (k) can be: H (high) =1-day; M (medium)=from 2 to 30 days; L (low) over 1 month;
- Effect (j) is the nature of the news/event. It can be B=bad; G=good; IL=in line.

The two macro area events returned by the NIM are general key events and specific key events. General key events are the events that indirectly influence the behavior of Eni, while specific key events are the events connected directly to the security.

Measuring the impact of news reveals that only a small percentage of news really has importance for the stock market. This percentage is almost equally divided among related and non-related news, but non-related news has a deeper on the stock price (Barazzetti et al. 2014).

11.2.2 Natural Language Processing and NewsVoc

We make use of a series of software tools for NLP whose aim is to help us to generate the behavioral schema of sentences related to events and to the NIM of each sentence itself: at first, we use a word segmentation tool, followed by a sense disambiguation process to identify the correct meaning of the sentences. Finally, we use a syntactical tool for the lemmatization of the text and a dictionary of synonyms to normalize it.

Once the data is normalized, we apply tags to the metadata that univocally identify the concept relative to specific values of NIM. NLP is applied to the news to extract the knowledge model: this is a semantic network of concepts that we have called NewsVoc that consists of the association between the values of NIM with the single event related to the news. The result is a set of tagged sentences, representing events, that are correlated by the same effects as measured by NIM, where, i.e., a= "tag_african_country"; b= "tag_rebellion"; d= "tag_plant_danger," and so on (see example in Table 11.2).

Table 11.2 NewsVoc model

Id_NewsVoc	NewsVoc_Concepts							NIM
1	*a*	*b*	*d*	F	H			+4
2	f	g	j	k				−1
3	s	d	c	v	N	m	n	+2
4	a	g	k	h	p			+1

Let's try to clarify this point with an example. Imagine that news 325 is "*More than 100 people are killed when religious violence flared in mainly Muslim towns in the north and in the southern city of Onitsha, Nigeria.*" This news is composed of +4 NIM. From the point of view of semantic network, we must link News 325 with some concept related to (i) the fact that we are talking about an African Country (coming from *Nigeria* lemma), and (ii) and we are talking about some kind of "rebellion" and "damage to the industrial plant" (coming from *Onitsha*). When we link the lemma coming from News 325 (high NIM, +4) with concept into the NewsVoc, we obtain the information that also the concepts *a, b*, and *d* are related with an high NIM (+4). NewsVoc allows us to disengage the NIM by the level of lemmas to that of a semantic to a higher level.

NLP is applied on each single news event daily queried by the user. We obtain metadata that is ready to be processed in the decision tree to establish at which vector of the NewsVoc it belongs. At this point, the NIM is attributed.

The combination of the NIM of different news events allows the Financial Prompter, a well-defined algorithm, to identify the future trend of the stock.

11.2.3 Financial Prompter

Financial Prompter (FP) is an integrated event-forecasting and trading-decision support system.

The inputs of the FP process are the daily news: they are processed through the NLP as described above to establish the correct relation between NewsVoc_concept in the NewsVoc model and the corresponding NIM value. The trend is determined by the following formula where *dfp* means "density of future probability":

$$dfp_{te} = \sum_{i}^{n} nim_i * c_i \tag{11.1}$$

where $t_e \mid \{t_e = 1; 1 < t_e \leq 30;\ t_e > 30\}$ expressed in days, n is the number of news/events at time t expressed in days, *nim* identifies the News Index Map as described above for that news, and c is the weight of each *nim*.

In this system, the user can choose from five trading rules:

Rule I: if $dfp_{te} >> 0 =>$ then the current trading strategy is "*strong buy*"
Rule II: if $dfp_{te} > 0 =>$ then the current trading strategy is "*buy*"
Rule III: if $dfp_{te} = 0 =>$ then the current trading strategy is "*hold*"
Rule IV: if $dfp_{te} < 0 =>$ then the current trading strategy is "*sell*"
Rule V: if $dfp_{te} << 0 =>$ then the current trading strategy is "*strong sell.*"

11.3 Data Analysis

We show the 2014 and 2015 results for Eni[3], an integrated energy company, listed in the FtseMib[4], belonging to energy sector. More precisely, the full sample consists of news and one-minute closing prices of Eni from December 31, 2013, through March 07, 2015.

The intraday financial data of the stock were taken from Bloomberg, while we obtained a collection of more than 15,000 financial and non-financial news events from *Il Sole 24 Ore* and *BorsaInside*[5].

From this pool of articles we discarded all news that could have an ambiguous effect (i.e., gossip or sport articles). After discarding all the news that were not relevant for our study, we were left with a total of 3877 financial and non-financial news articles correlated with Eni. Then we applied the NewsVoc model to the news after a preprocess of normalization of texts to extract vector of words. Finally. we forecasted the trend of Eni stock by applying the Financial Prompter on the output of NewsVoc.

11.4 Empirical Results

The predictive model, applied to the news of the year 2012 and 2013 as a back-test, is now working and the very first results encouraged us to deeply extend the time frame of the prediction.

The potential predictive power of NewsMarket is illustrated in Figs. 11.2, 11.3, and 11.4. Applying the system to a specific stock or sector led to more accurate predictions of price direction. It was reasoned that keywords specific to the company firstly and related to the sector secondly were more influential in determining price direction.

In all the three time frames, the Financial Prompter *dfp* had the power to suggest the direction of the trend several days or even months in advance.

We examined the *dfp* signal for the different time frames. In Fig. 11.2, we depict the results of *dfp* (left scale) and the price of Eni (right scale) between January 2012 and December 2014 for six semesters.

It can be easily seen that there is a good visual correlation between the *dfp* and the Eni price. Another important thing to highlight is that there is a "lag" between the two values. For example, the *dfp* signal in point 7 suggests the stock trend for the next semester (January–June 2015). This indicates that *dfp* expressed a positive sentiment before the stock started rising.

[3] Active in more than 70 countries, Eni operates in the oil and gas, electricity generation and sale, petrochemicals, oilfield services construction and engineering industries.

[4] The FTSE MIB Index is the primary benchmark index for the Italian equity market.

[5] Although the trading starts at 9:00 A.M. and closes at 17:30 CET we felt important to consider news article releases during all the day (also news posted after closing hours) and for every day (including weekends and holidays).

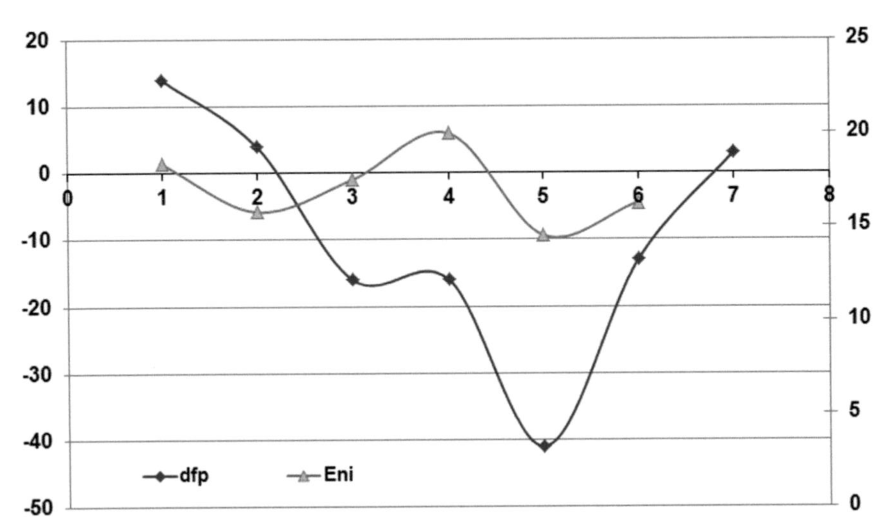

Fig. 11.2 Long time frame (*semester*). Simulated trend elaborated by the Financial Prompter (*left scale*) compared with Eni price (*right scale*) from January 2012 to December 2014

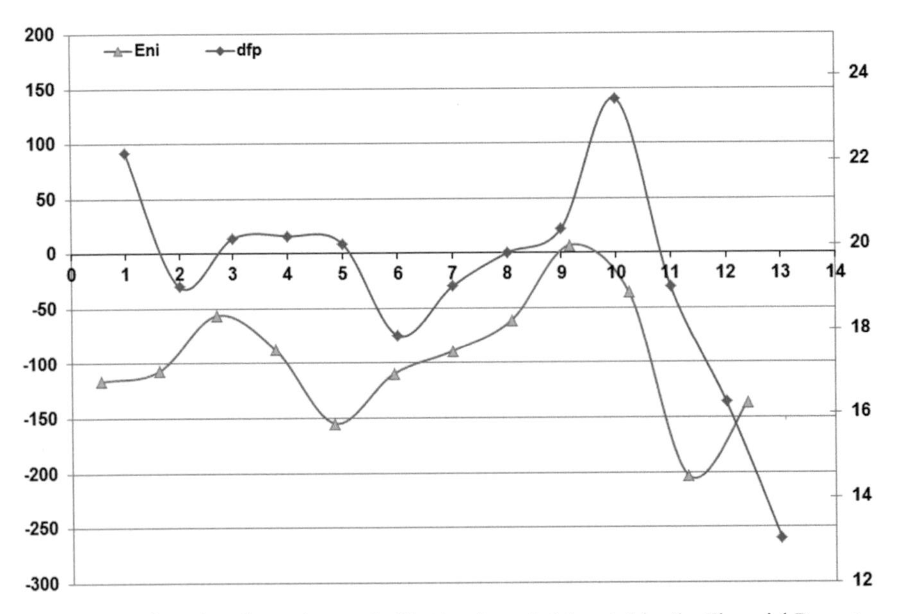

Fig. 11.3 Medium time frame (*quarter*). Simulated trend elaborated by the Financial Prompter (*left scale*) compared with Eni price (*right scale*) from January 2012 to December 2014

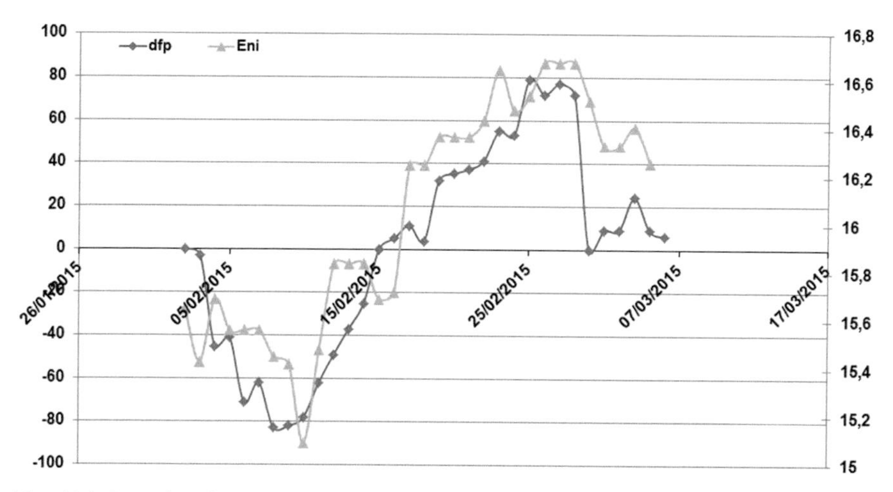

Fig. 11.4 Low time frame (*daily*). Simulated trend elaborated by the Financial Prompter (*left scale*) compared with Eni price (*right scale*) from January 2015 to March 7, 2015

Table 11.3 Sign of the trend concerning the long time frame

Period	Semester	Trend
1	I Sem- 2012	Positive
2	II Sem- 2012	Negative
3	I Sem- 2013	Negative
4	II Sem- 2013	Negative
5	I Sem- 2014	Negative
6	II Sem- 2014	Positive
7	I Sem- 2015	Positive

Table 11.3 summarizes all the *dfp* trend signals for each semester. Expectations for signals related to the third and fourth semesters for all the other signals are in line with the trend.

In Fig. 11.3, we present the results from January 2012 to the first quarter of 2015 (until March 7th). The time frame comprises 13 quarters. Unlike the semester's signal, the next two time frames were more accurate in defining the trend of Eni. There is a strong visual correlation between the *dfp* and the Eni price with respect to the previous time frame.

Our method can thus clearly detect the direction of market sentiment, which is closely related to the direction of the actual stock price movement.

Figure 11.4 shows the daily trend from January 2015 to March 7, 2015.

Looking at the three time frames we considered, we observed that the medium time frame and the low time frame (Figs. 11.3 and 11.4) were more accurate in predicting the stock-price trends of Eni (Table 11.4).

The value of the *dfp* depends on how long negative or positive events, defined in our NewsVoc, persist. A specific event, for example, may influence the *dfp* more, whereas a general event may not. One might expect, therefore, some variation in signal strength to be accounted for by variation in the trends.

Table 11.4 Sign of the trend concerning the medium time frame

Period	Quarter	Trend
1	I Quart- 2012	Positive
2	II Quart - 2012	Negative
3	III Quart - 2012	Positive
4	IV Quart - 2012	Positive
5	I Quart - 2013	Negative
6	II Quart - 2013	Negative
7	III Quart - 2013	Positive
8	IV Quart - 2013	Positive

11.5 Conclusions

Can linguistic information, compiled with the aid of human agents, help predict financial trends/economic activity?

Web news can be used to accurately track not only social phenomena but also financial trends. Investors, economists, and journalists follow monthly macroeconomic data releases on economic conditions. There is a large literature about this topic, i.e., how macroeconomic news influenced markets. A huge problem is that this information is available with a time lag: are generally released about halfway through the next month and are typically revised several months later.

By analyzing daily news events and correlating them with the current level of economic activity with NIM, NewsMarket is able to help investors and economists to suggest trends in a specific sector for the next month. NewsMarket is not a "crystal ball" that predict the future, but it is a tool to obtain early indications of movements in the financial markets. The tool serve as a baseline to help analysts and investors get started with their own modeling efforts that can subsequently be refined for specific applications.

As we have demonstrated, NewsMarket's best (i.e., most accurate) time frame is the quarterly period. In fact, by crawling online news every day, NewsMarket has the capability to identify "turning points" three months in advance. These findings raise interesting questions regarding the circumstances under which online-news\-based predictions might be useful.

Finally, we note two further points that suggest the potential value of online-news- based predictions. First, modest performance gains may still prove useful for applications, not only for financial analysis, but also for other domains such as policymakers and supervising committees, where even a minimal performance edge can be valuable. Second, unlike other data sources that require customized and often cumbersome collection strategies, online news can be collected for many domains simultaneously and easily analyzed along geographic and other dimensions, all in real time.

As the product evolves every day, we expect to obtain more accurate estimation of the financial trends also in other time frame. At the moment we are working to create a dedicated events database for sectors other than oil and gas.

References

Barazzetti, A., Cecconi, F., & Mastronardi, R. (2014). Financial forecasts based on analysis of textual news sources: Some empirical evidence. *Artificial economics and self organization - Agent-Based Approaches to Economics and Social Systems, 669,* 133–145. New York: Springer.

Bouchaud, J.-P. (2009). The (unfortunate) complexity of the economy. *Physics World, 4,* 28–32.

Casti, J. (2012). Eventi X, Eventi estremi e il futuro della civiltà. Il Saggiatore.

Choi, H., & Varian, H. (2009) Predicting the present with Google trends. Technical report.

Grimmer, J., & Stewart, B. M. (2013). Text as data: The promise and pitfalls of automatic content 383 analysis methods for political texts. *Political Analysis, 21*(3), 267–297.

Jonathan, C., Boyd-Graber, J., Wang, C., Gerrish, S., & Blei, D. M. (2009). Reading tea leaves: How humans interpret topic models. In Y. Bengio, D. Schuurmans, J. Lafferty, C. K. I. Williams, & A. Culotta (Eds.), *Advances in neural information processing systems* (pp. 288–296). Cambridge: The MIT Press.

Mittermayer, M. (2004). Forecasting intraday stock price trends with text mining techniques. Hawaii International Conference on System Sciences, Kailua-Kona, HI.

Pestov, V. (2013). Is the -NN classifier in high dimensions affected by the curse of dimensionality? *Computers and Mathematics with Applications, 65,* 1427–1437.

Preis, T., Moat, H. S., & Stanley, H. E. (2013). Quantifying trading behavior in financial markets using Google Trends. *Scientific Report, 3,* 1684.

Scheffer, M. (2009). *Critical transitions in nature and society*. Princeton: Princeton University Press (Princeton Studies in Complexity)

Schumaker, R. P., & Chen, H. (2006). Textual analysis of stock market prediction using financial news articles. 12th Americas Conference on Information Systems (AMCIS-2006), Acapulco, Mexico.

Soni, A., van Eck, N. J., & Kaymak, U. (2007). Prediction of stock price movements based on concept map information. IEEE symposium on computational intelligence in multicriteria decision making (pp. 205–211).

Xie, L., Sundaram, H., & Campbell, M. (2008). Event mining in multimedia streams. Vol. 96, No. 4, April 2008. Proceedings of the IEEE.

Index

F. Cecconi (ed.), *New Frontiers in the Study of Social Phenomena*,
DOI 10.1007/978-3-319-23938-5

Printed in the United States
By Bookmasters